Representing 3-Manifolds
by Filling Dehn Surfaces

K&E Series on Knots and Everything — Vol. 58

Representing 3-Manifolds by Filling Dehn Surfaces

Rubén Vigara
Álvaro Lozano-Rojo
Centro Universitario de la Defensa - IUMA, Spain

World Scientific

NEW JERSEY · LONDON · SINGAPORE · BEIJING · SHANGHAI · HONG KONG · TAIPEI · CHENNAI · TOKYO

Published by

World Scientific Publishing Co. Pte. Ltd.

5 Toh Tuck Link, Singapore 596224

USA office: 27 Warren Street, Suite 401-402, Hackensack, NJ 07601

UK office: 57 Shelton Street, Covent Garden, London WC2H 9HE

Library of Congress Cataloging-in-Publication Data

Names: Vigara, Rubén. | Lozano-Rojo, Álvaro.

Title: Representing 3-manifolds by filling Dehn surfaces / Rubén Vigara
   (Centro Universitario de la Defensa--IUMA, Spain), Álvaro Lozano-Rojo
   (Centro Universitario de la Defensa--IUMA, Spain).

Description: New Jersey : World Scientific, 2016. | Series: Series on knots and everything ; vol. 58 |
   Includes bibliographical references and index.

Identifiers: LCCN 2016002805 | ISBN 9789814725484

Subjects: LCSH: Dehn surgery (Topology) | Three-manifolds (Topology) | Topological manifolds.

Classification: LCC QA613.658 .V54 2016 | DDC 514/.34--dc23

LC record available athttp://lccn.loc.gov/2016002805

British Library Cataloguing-in-Publication Data

A catalogue record for this book is available from the British Library.

Copyright © 2016 by World Scientific Publishing Co. Pte. Ltd.

All rights reserved. This book, or parts thereof, may not be reproduced in any form or by any means, electronic or mechanical, including photocopying, recording or any information storage and retrieval system now known or to be invented, without written permission from the publisher.

For photocopying of material in this volume, please pay a copying fee through the Copyright Clearance Center, Inc., 222 Rosewood Drive, Danvers, MA 01923, USA. In this case permission to photocopy is not required from the publisher.

Printed in Singapore

to Yolanda, Julia and Pablo
RV

to Silvia and Yago
AL

Preface

The term *Dehn surface* was introduced in the celebrated work [51] where Papakyriakopoulos proved Dehn's Lemma, together with the Loop and Sphere Theorems. In fact, the expression *Dehn disk* is a simplification of

disk to which Dehn's Lemma applies.

A Dehn surface Σ in a closed 3-manifold M is a compact immersed surface with only double point (Fig. I(a)) or triple point (Fig. I(b)) singularities. If Σ has boundary, the singularities of Σ are required to be disjoint from the boundary. If $f : S \to M$ is an immersion such that $f(S) = \Sigma$, where S is a compact surface, the inverse images under f of the double and triple points of Σ, together with the information about how they become identified by f form the *Johansson diagram* of Σ [36, 37] (Fig. II). The Johansson diagram is often called *double decker set* by 2-knot theorists (cf. [11]).

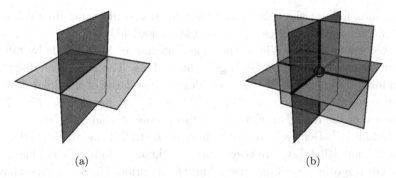

(a)   (b)

Fig. I: Singularities of a Dehn surface.

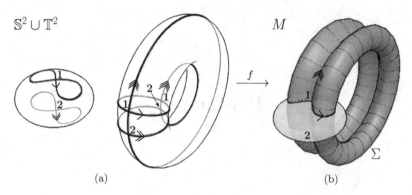

Fig. II: A Dehn surface and its Johansson diagram.

The Dehn surface Σ *fills* M if it defines a cellular decomposition of M in which the 0-skeleton is the set of triple points of Σ, the 1-skeleton is the set of singularities of Σ, and the 2-skeleton is Σ itself. In a short note at the end of [24], Haken asserts that every homotopy 3-sphere has a filling Dehn sphere, and he proposes filling Dehn spheres and their Johansson diagrams as a method for representing homotopy 3-spheres. In the same work, he gives a sufficient condition for a Johasson diagram to represent a simply-connected 3-manifold. Haken's ideas were not investigated until [49], where Montesinos proves that every closed orientable 3-manifold has a filling Dehn sphere. Before that, a slightly weaker result is proved by Fenn and Rourke in [16]: every 3-manifold has a Dehn sphere Σ such that $M - \Sigma$ is a disjoint union of open 3-balls (these spheres are called *quasi-filling* in [4]). The key idea is the following:

if Σ fills M, it is possible to construct M from the Johansson diagram of Σ.

Therefore, filling Dehn spheres and their Johansson diagrams are a suitable way for representing all closed orientable 3-manifold.

According to Montesinos, the representation of 3-manifolds by filling Dehn spheres seems to be "stronger" than others. The Johansson diagram of a filling Dehn sphere Σ of M provides the 2-skeleton of a cellular decomposition of M and an immersion $f : \mathbb{S}^2 \to M$ that parametrizes Σ. The presence of the map f is what makes this representation different.

Before [49], Dehn spheres in $\mathbb{R}^3$ have been studied due to their relation with 2-knots [10,18], where they appear as "knot projections" of 2-knots in $\mathbb{R}^4$. On the other hand, generic filling Dehn surfaces in 3-manifolds have been also studied as the canonical immersed surfaces of cubed manifolds [2].

Nevertheless, the viewpoints of those works is quite different from that of Haken and Montesinos. After [49], some research on this topic has been made by the authors [42–44, 67–70], and by Amendola [3–5].

This book is an up-to-date English translation of the first author's PhD thesis [68], together with an overview of the results included in all these works. Since this book is inspired by Haken-Montesinos viewpoint, it mainly deals with filling Dehn spheres, although more general filling Dehn surfaces are also taken into account.

The main result of the book (Theorem 5.8) is stated in Chap. 5. It is a Reidemeister-type theorem for filling Dehn spheres that explains how different filling Dehn spheres of the same 3-manifold are related to each other. The most natural way to deform an immersion is by regular homotopy. A regular homotopy can be decomposed into a finite set of moves (the *Homma-Nagase moves* [32, 33], see also [26, 57]), together with ambient isotopies. We slightly modify the Homma-Nagase moves, introducing an equivalent set of moves (*Haken moves*) better suited to the study of filling immersions. A Haken move applied to a filling Dehn surface is an *f-move* (Fig. III) if the resulting surface is still filling. Two filling Dehn surfaces are said to be *f-homotopic* if they are related by a finite sequence of *f*-moves and ambient isotopies. On the other hand, for immersions of the 2-sphere into 3-manifolds, *regularly homotopic* is equivalent to *homotopic* [26, 41, 63]. In particular, null-homotopic filling Dehn spheres of the same 3-manifold are regularly homotopic. Our main theorem for immersions can be stated as follows:

Main Theorem 1 (Theorem 5.8). *Null-homotopic filling Dehn spheres of the same 3-manifold are f-homotopic.*

Recall that for irreducible 3-manifolds, all the Dehn spheres are null-homotopic. The set of *f*-moves and Theorem 5.8 have natural translations to Johansson diagrams that are also explained in Chap. 5, and that lead to the notion of *f-equivalency* of filling Johansson diagrams. A filling Johansson diagram in $\mathbb{S}^2$ is *null-homotopic* if it represents a null-homotopic filling Dehn sphere. Although we know no algorithm to decide if a filling Johansson diagram in $\mathbb{S}^2$ is null-homotopic, [69] introduces an algorithm (*duplication*) that for any filling Johansson diagram $\mathcal{D}$ in $\mathbb{S}^2$ gives another filling diagram $\mathrm{dup}(\mathcal{D})$ (the *duplicate* of $\mathcal{D}$) that is null-homotopic and represents the same 3-manifold as $\mathcal{D}$. This algorithm is detailed in Sec. 5.5. Out main theorem for diagrams is also stated in this section:

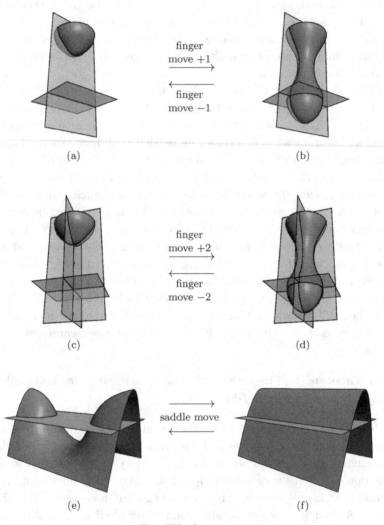

Fig. III: f-moves.

Main Theorem 2 (Theorem 5.26). *Two filling diagrams in* $\mathbb{S}^2$ *represent the same 3-manifold if and only if their duplicates are f-equivalent.*

Although our set of f-moves is quite simple and natural, it has one disadvantage: finger moves -1 and saddle moves (Fig. III) are not always f-moves. In other words, if we apply a finger move -1 or a saddle move to a filling Dehn surface, the resulting surface could be non-filling. This problem is solved in [3] by defining an equivalent set of moves for filling Dehn surfaces (*Amendola's moves*), briefly presented in Sec. 5.6. It turns out that two filling Dehn surfaces are f-homotopic if and only if they are related by Amendola's moves, and that the resulting surface after an Amendola's move is always filling. Using this property, Amendola's moves are used to define a new 3-manifold invariant [3] using Turaev-Viro state sum machinery [65].

Most of the book is devoted to the proof of Theorem 5.8, which is quite long and tedious. That is why it has been postponed to a separate chapter (Chap. 6). It makes use of a wide variety of techniques and results from differential topology [20, 30, 40, 64, 66]; regular homotopies and isotopies [26, 32, 33, 35, 60]; triangulations, subdivisions and simplicial collapsings [8, 14, 71, 72]; and combinatorics (shellability) [8, 9, 60]. This is the first time that the complete proof of Theorem 5.8 is published in English. The Spanish version of this proof appeared in [68], and an outline of this proof appears in [69].

The rest of the book is organized as follows.

Chapter 1 introduces general concepts and notations that will be used throughout the rest of the book. Chapter 2 is devoted to the basic definitions about Dehn surfaces and filling Dehn surfaces.

Every Dehn surface has a Johansson diagram. On the other hand, we can define *diagram* in an abstract sense as a collection of coherently identified curves on a surface, and it is natural to ask if a given diagram of this type is *realizable*: if it is the Johansson diagram of some Dehn surface in some 3-manifold. Johansson solved this question in [36] for generic (orientable or non-orientable) 3-manifolds, and in [37] for orientable 3-manifolds. In each case necessary and sufficiente conditions for a given diagram to be realizable were given. Although [36] and [37] were written in the context of the search of a correct proof of Dehn's Lemma, so they deal only with Dehn disks, their techniques apply also for diagrams in compact connected surfaces. Since we restrict our attention to the orientable case, we only focus in the results of [37]. The techniques of [37] can also be used to decide if a given diagram $\mathcal{D}$ in a closed orientable surface is *filling*, i.e., if it is the

Johansson diagram of a filling Dehn surface of a 3-manifold M and, if this is the case, to recover M from $\mathcal{D}$. All these topics are discussed in Chap. 3.

Chapter 4 starts with a study of the coverings of a Dehn sphere. An n-sheeted covering of a Dehn sphere defines, in a quite natural way, a map from the set of curves of the Johansson diagram to the group of permutations of n elements. A detailed analysis of this map allows to find a presentation of the fundamental group of the Dehn sphere from its Johansson diagram. This presentation was already known, probably by Haken (see [39, Problem 3.98]), but we had not seen it in printed form until [69]. A generalization of this presentation to a wider class of Dehn surfaces has been obtained in [42], and it is stated here for completeness. Since the fundamental group of a 3-manifold is isomorphic to the fundamental group of any of its filling Dehn surfaces, those presentations can be used to study 3-manifold groups.

A Dehn sphere always has an even number of triple points [24]. Therefore the minimal number of triple points of a filling Dehn sphere is 2. The Dehn spheres in $\mathbb{S}^3$ with 2 triple points are classified in [62] (for immersions the projective plane with one triple point see [21]). It turns out from [62] that there are only three different filling Dehn spheres in $\mathbb{S}^3$ with 2 triple points (*Shima's spheres*). In Chap. 7 it is shown how the Johansson diagrams of these three examples can be transformed into each other using f-moves, and we present also other two examples of filling Dehn spheres with two triple points: one in $\mathbb{S}^2 \times \mathbb{S}^1$ and the other in the lens space $L(3,1)$. These are the only examples of filling Dehn spheres with two triple points (Theorem 7.2). Some other questions about minimal numbers of triple points are discussed in Chap. 7.

A filling Dehn sphere Σ in a 3-manifold M is *minimal* [69] if there is no filling Dehn sphere in M with less triple points than Σ. The number of triple points of a minimal filling Dehn sphere of M is the *Montesinos complexity* of M [43]. Then, Theorem 7.2 classifies the 3-manifolds with Montesinos complexity 2. Some other results about the Montesinos complexity [43] are stated: a sharp bound for the Montesinos complexity of a connected sum of 3-manifolds and some results about 3-manifolds with Montesinos complexity 4. The concept of minimal filling Dehn sphere easily generalizes to more general surfaces, so we can talk about minimal filling Dehn tori or minimal genus g filling Dehn surfaces (*filling Dehn g-tori* in our notation). The number $t_g(M)$ of triple points of a minimal filling genus $g \geq 0$ Dehn surface of M is the *genus g triple point number* of M ($t_0(M)$ is the Montesinos complexity), and the ordered sequence

$$\mathcal{T}(M) = \big(t_0(M), t_1(M), t_2(M), \ldots\big)$$

is the *triple point spectrum* of the 3-manifold M. All these 3-manifold invariants were defined in [69]. They give a measure of the complexity of a 3-manifold in a similar way as the Heegaard genus, for example. These invariants are pretty hard to compute. We present some results about them from [42], as the only known triple point spectra ($\mathcal{T}(\mathbb{S}^3)$ and $\mathcal{T}(\mathbb{S}^2 \times \mathbb{S}^1)$), for example. A closely related 3-manifold invariant is the *surface-complexity* of M, defined in [4] as the minimal number of triple points of a quasi-filling Dehn surface in M. For completeness, Sec. 7.5 includes some results from [4,5] about surface-complexity.

In the final chapter (Chap. 8) we discuss some ideas that can relate filling Dehn spheres with 2-knots [70] and 1-knots [44]. We finish with a list of open problems that could be interesting for further research.

For a better reading of the proof of Theorem 5.8, we have left to the appendix three long proofs of partial results stated in Chap. 6.

This book is intended as a basic reference book for researchers in low-dimensional topology. Nevertheless, it can also be used as an advanced textbook for graduate students or even for advanced undergraduates in mathematics interested in learning techniques of geometric topology and the fundamental tools in the study of Dehn surfaces in 3-manifolds.

In the better tradition of low-dimensional topology books, it has hundreds of pictures most of them 3D pictures, that try to explain and illustrate as much as possible all the concepts and ideas introduced in the text.

Acknowledgements

There are many people whose support helped this book to come to light.

Both authors are very grateful to J.M. Montesinos, M.T. Lozano and M. Avendano for their valuable suggestions and comments; to their colleagues of the research group "Geometría" at IUMA-Universidad de Zaragoza; to Centro Universitario de la Defensa (CUD) de Zaragoza and its Director, A. Elipe, for their invaluable support; and to the editor Zhang Ji for their patience and support during the writing of this book.

First author wants to acknowledge his colleagues at the Department of Fundamental Mathematics of the National Distance Education University (UNED, Spain), at the Department of Mathematics of the Carlos III University of Madrid, at IES Antonio Machado (Madrid) and at CUD de Zaragoza, with special mention to A.F. Costa, F. Terán, and A. Pérez. He is also very grateful to the second author: without his help this book would have stayed forever as a to-do project.

The second author is indebted to many others, like the his colleagues and friends at the Department of Mathematics of the University of the Basque Country, at the Department of Geometry and Topology from the University of Santiago de Compostela and at the Department of Algebra, Geometry and Topology from the University of Valladolid, specially to F. Alcalde, F. Cano, J. González, P. González-Sequeiros, R. Ibañez and M. Macho.

Last but not least, he also want to thank the first author for introducing him into the beautiful topic of Dehn surfaces and for his generosity through the works they have started together.

Both authors have been supported by the European Social Fund and Diputación General de Aragón (Grant E15 Geometría), and by Spanish Government's *Programa Estatal de Fomento de la Investigación Científica y Técnica de Excelencia* (research projects MTM2013-45710-C2 and MTM2013-46337-C2 for the first and second author respectively).

This book contains material from the first author's PhD thesis [68], that was supported by a predoctoral grant from UNED.

RV & AL
Zaragoza, 2016

Contents

Chapter 1

Preliminaries

1.1 Sets

For a given a map $f : A \to B$ between two sets, the sets A, B and $f(A) \subset B$ are called the *domain*, *codomain* and *image* of f, respectively. The *singular points* or *singularities* of f are the elements of B with more than one preimage under f, and the *pre-singular points* of f are the preimages under f of the singularities of f. The set of all singularities of f is the *singular set* of f, denoted by $\mathrm{Sing}(f) \subset B$, and the set of pre-singular points of f is the *pre-singular set* of f, denoted by pre-$\mathrm{Sing}(f) \subset A$. Obviously, $f\big(\text{pre-}\mathrm{Sing}(f)\big) = \mathrm{Sing}(f)$.

Given $y \in \mathrm{Sing}(f)$, *the multiplicity of y with respect to f* is the number of preimages of y under f, that is, the cardinality of the set $f^{-1}(y)$. An element $y \in B$ is a *double point, triple point, quadruple point, etcetera* if its multiplicity is 2, 3, 4, *etcetera*, respectively.

Two elements $x, x' \in A$ are *related by f* if $f(x) = f(x')$.

The *union of two maps* $f : A \to C$ and $g : B \to C$ is the map

$$f \cup g : A \sqcup B \to C$$

whose domain is the disjoint union of the domains of f and g defined in the natural way. If $A = B$, $A \sqcup A$ is the disjoint union of two different copies of A. The union of $n \geq 3$ maps with the same codomain is defined similarly.

Two maps $f, g : A \to B$ *agree over* $D \subset A$ if $f|_D = g|_D$.

1.2 Manifolds

In the present book, unless otherwise stated, all manifolds and maps are considered smooth (C^∞), all 3-manifolds are closed (compact, connected and without boundary) and orientable, and all 2-manifolds (*surfaces* from now

1

on) and 1-manifolds are compact and orientable with possibly non-empty boundary.

Any compact connected manifold is metrizable and separable. The theorem of Radó [55] asserts that all 2-manifolds are triangulable. The same holds for 3-manifolds [48]. It is also well-known that all n-manifolds, with $n \leq 3$, admit a unique differentiable structure [74]. Since most of the manifolds that are considered in this book are 2- or 3-manifolds, there is no need to specify their differentiable structure. For further details on manifolds see [22, 30].

Given a manifold N, ∂N denotes the boundary of N (which might be empty) and $\text{int}(N) = N - \partial N$ denotes its interior. As usual, TN is the tangent bundle of N and $T_P N$ is the tangent space of N at $P \in N$. If N' is another manifold and $f : N \to N'$ is a smooth map, $df : TN \to TN'$ is the *derivative of f* and $df_P : T_P N \to T_{f(P)} N'$ is the restriction of df to $T_P N$. The map f is called *normal* if pre-Sing(f) does not meet ∂N, and f is called an *immersion* if df_P is injective for every $P \in N$. The Implicit Function Theorem implies that every immersion is a local diffeomorphism. An immersion is an *embedding* if it is a homeomorphism between its domain and its image.

For $n \geq 1$, $\mathbb{S}^n$ and $\mathbb{D}^n$ denote the n-dimensional sphere and closed ball (also called the n-disk), respectively. As usual, $\partial \mathbb{D}^n = \mathbb{S}^{n-1}$. A closed orientable surface of genus $g \geq 0$ is often called a *g-torus*. Usually S and M denote a generic surface and a generic 3-manifold respectively. Unless otherwise stated, I denotes the unit interval $[0, 1]$. Of course, in most definitions and constructions involving I, this interval can be replaced by an arbitrary interval obtaining equivalent objects.

1.3   Curves

A *path* in a manifold N is a map from I into N. Let $\alpha : I \to N$ be a path in N. It is said that α is a path *from $\alpha(0)$ to $\alpha(1)$*, and the points $\alpha(0)$ and $\alpha(1)$ are called *endpoints* of α. The path α is *closed* if its endpoints coincide. The *inverse path* of α is the path α^{-1} defined by $\alpha^{-1}(t) = \alpha(1 - t)$ with $t \in I$. If β is another path in N with $\alpha(1) = \beta(0)$, the *concatenation of α and β*, is the path

$$\alpha * \beta(t) = \begin{cases} \alpha(2t) & \text{if } t \leq 1/2, \\ \beta(2t - 1) & \text{if } t \geq 1/2. \end{cases}$$

Fig. 1.1: A multiple point.

Replacing $2t$ by an appropriate bump function $\phi(t)$ with $\phi(0) = 0$ and $\phi(1/2) = 1$ [30, p. 42], it can be assumed that $\alpha * \beta$ is smooth.

The image of a path in N which is an embedding is an *arc* in N.

A *parametrized closed curve* in a manifold N is an immersion $\alpha : \mathbb{S}^1 \to N$, and its image $\alpha(\mathbb{S}^1)$ is a *closed curve*. If α is a path or a parametrized closed curve in N, its image $\alpha(I)$ or $\alpha(\mathbb{S}^1)$, respectively, is denoted $|\alpha|$. In either case α is called a *parametrization* of $|\alpha|$. For simplicity, the term "parametrized" is omitted when no confusion can arise.

Given a set of paths or parametrized closed curves $\mathcal{C} = \{\alpha_1, \ldots, \alpha_n\}$ in N, we denote $|\mathcal{C}| = |\alpha_1| \cup \cdots \cup |\alpha_n|$.

Remark 1.1. Since both closed curves and arcs are images of immersions, neither of them have "corners". The only allowed corner-like points are the singular points of parametrized closed curves (see Fig. 1.1).

1.4 Transversality

In this book transversality appears in the study of immersions of surfaces into 3-manifolds and immersions of curves into surfaces. A normal immersion $f : N \to N'$ of codimension 1 (i.e., $\dim N' = \dim N + 1$) is *transverse* if:

(1) the multiplicity with respect to f of any point of N' is at most $\dim N'$; and

(2) if $P_1, \ldots, P_k \in N$ are different points of N related by f and $\bar{P}$ is their common image under f, the vector subspace $\bigcap_{i=1}^{k} df_{P_i}(T_{P_i}N)$ of $T_{\bar{P}}N'$ has codimension k.

Let S be a surface (connected or not), let M be a 3-manifold and let $f : S \to M$ be a transverse immersion.

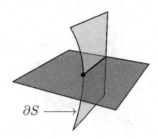

$\partial S \longrightarrow$

Fig. 1.2: A singular point in $f(\partial S)$.

Since f is transverse, the multiplicity of a singular point is at most 3 and there are no tangencies. Since f is normal, there is no singular point in $f(\partial S)$ like the one shown in Fig. 1.2. Therefore, if P is a singular point of f, then P must be a double or a triple point of f, and there is a neighbourhood of P like the one in Fig. 1.3(a) or Fig. 1.3(b) depending on whether P is a double or triple point.

Finally, a collection $\mathcal{C} = \{\alpha_1, \ldots, \alpha_n\}$ of parametrized closed curves in a surface S form a *normal system of curves* of S [51] if the immersion $\alpha_{\mathcal{C}} = \alpha_1 \cup \cdots \cup \alpha_n$ is transverse, that is:

(1) at most two branches of curves of $\mathcal{C}$ meet at a point of S; and
(2) if P is a singular point of $\alpha_{\mathcal{C}}$, the two branches of curves of $\mathcal{C}$ meeting at P are transverse at P.

In general, we reserve the name *double point* for immersions of surfaces into 3-manifolds, and we use instead the term *crossing* for a double point of a normal system of curves in a surface.

1.5   Regular deformations

A *homotopy* is a map $H : N \times I \to N'$, where N and N' are manifolds. In this case, for each $t \in I$, $H_t : N \to N'$ is the map given by $H_t(P) = H(P, t)$. The map H is said to be a homotopy *between H_0 and H_1*. Since H can be regarded as a path in the space of functions $C^\infty(N, N')$, it is said that H *connects H_0 and H_1*. The homotopy H is *regular* if H_t is an immersion for every $t \in I$, and it is an *isotopy* if H_t is an embedding for every $t \in I$. Two maps $f, g : N \to N'$ are *homotopic* (resp. *regularly homotopic*, resp. *isotopic*) if there exists a homotopy (resp. regular homotopy, resp.

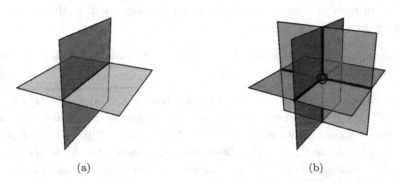

(a) (b)

Fig. 1.3: Double and triple points of a transversely immersed surface.

isotopy) $H : N \times I \to N'$ connecting them. Given two homotopies (resp. regularly homotopies, isotopies) $H, F : N \times I \to N'$ such that $H_1 = F_0$, the *concatenation of H and F*, defined by

$$H * F(P,t) = \begin{cases} H(P,2t) & \text{if } t \le \frac{1}{2}, \\ F(P,2t-1) & \text{if } t \ge \frac{1}{2}, \end{cases}$$

proves that being homotopic (resp. regularly homotopic, isotopic) is an equivalence relation. In the same way as in the concatenation of paths, it can be assumed that $H * F$ is smooth. A map $f : N \to N'$ is *null-homotopic* if it is homotopic to a constant map.

Given the homotopy $H : N \times I \to N'$, the *track of H* [30, p. 178] is the level preserving map $\widetilde{H} : N \times I \to N' \times I$ given by

$$\widetilde{H}(P,t) = \big(H(P,t),t\big).$$

An *ambient isotopy* of a manifold N is a map $\zeta : N \times I \to N$ such that ζ_t is a diffeomorphism for all $t \in I$ and $\zeta_0 = \text{id}_N$. Two immersions $f, g : N \to N'$ are *ambient isotopic in N* if there exists an ambient isotopy ζ of N such that $f \circ \zeta_1 = g$. Similarly, f and g are *ambient isotopic in N'* if there exists a ambient isotopy ζ' in N' with $\zeta'_1 \circ f = g$. A homotopy $H : N \times I \to N'$ is an *ambient isotopy* if there exist ambient isotopies ζ, ζ' of N, N' respectively, such that the diagram

$$\begin{array}{ccc} N & \xrightarrow{H_0} & N' \\ \zeta_t \uparrow & & \downarrow \zeta'_t \\ N & \xrightarrow{H_t} & N' \end{array}$$

commutes for all $t \in I$, that is, $H(P,t) = \zeta'\big(H_0(\zeta(P,t)),t\big)$ for all $P \in N$ and $t \in I$. The immersions $f, g : N \to N'$ are *ambient isotopic* if there exists an ambient isotopy connecting them. Using stability theory [20] it can be proved that:

Lemma 1.2. *A regular homotopy $H : N \times I \to N'$ such that H_t is transverse for all $t \in I$ is an ambient isotopy.* $\square$

A diffeomorphism $f : N \to N$ *fixes* a subset $K \subset N$ if $f(P) = P$ for all $P \in N$, that is, K is a set of *fixed points* of f. Similarly, a regular homotopy, isotopy or ambient isotopy $H : N \times I \to N'$ fixes $K \subset N$ if $H_t(P) = H_0(P)$ for all $P \in K$ and $t \in I$. In these cases, if a map fixes K, it also fixes $\mathrm{cl}(K)$. A diffeomorphism, regular homotopy, isotopy or ambient isotopy f is *supported* by $X \subset N$ if $N - X$ is fixed by f.

1.6   Complexes

Two common tools used throughout this book are cellular decompositions and triangulations of manifolds. For a gentle introduction to CW-complexes and basic results on cellular decompositions see [45,73]. The reader can find details on triangulations in [19,34,58].

Let N be a topological space. A *cellular decomposition K* of N is a collection of pairwise disjoint subsets of N, called *cells*, whose union is the manifold N and such that:

(1) A cell $\kappa \in K$ is either a point or it is homeomorphic to an open i-dimensional ball. In each case we say that κ is a 0-cell or an i-cell, respectively, and that the number (0 or i) is the *dimension* of κ.

(2) For each $i \in \mathbb{N}$, the *i-skeleton K^i* of K is the union of all cells of dimension lower or equal to i. For each $\kappa \in K$ of dimension $i \geq 1$, there exists a continuous map $\hat{q}_\kappa : \mathbb{D}^i \to N$, called *characteristic map*, such that $\hat{q}_\kappa(\partial \mathbb{D}^i) \subset K^{i-1}$ and $q_\kappa = \hat{q}_\kappa|_{\mathrm{int}(\mathbb{D}^i)} : \mathrm{int}(\mathbb{D}^i) \to \kappa$ is a homeomorphism.

The pair (N, K) is a *cellular complex*.

Remark 1.3. By the invariance of domain theorem, a cell decomposition of an n-manifold has no cells of dimension greater than n.

Remark 1.4. Cellular decomposition can be extended to finite and infinite dimensional spaces more general than manifolds (CW-complexes).

Let K be a cellular decomposition of N and let κ and $\kappa' \in K$ be two cells of K. If $\mathrm{cl}(\kappa) \subset \mathrm{cl}(\kappa')$, κ is called a *face* of κ' and this is denoted by

$\kappa < \kappa'$. The cells κ and κ' are *incident* if $\kappa < \kappa'$ or $\kappa' < \kappa$, and *adjacent* if $\mathrm{cl}(\kappa) \cap \mathrm{cl}(\kappa') \neq \emptyset$.

For a cell κ of K, the *star of κ in K*, written $\mathrm{star}_K(\kappa)$, is the union of all $\kappa' \in K$ such that $\kappa < \kappa'$. The *link* of κ in K is the set

$$\mathrm{link}_K(\kappa) = \mathrm{cl}\big(\mathrm{star}_K(\kappa)\big) - \mathrm{star}_K(\kappa).$$

If the characteristic map $\hat{q}_\kappa$ of a cell $\kappa \in K$ is an homeomorphism over its image, the cell κ is *regular*. If all cells of the complex K are regular, we say that K is *regular*. For more details on regular complexes see [45, p. 243].

A *subcomplex* of K is a closed subspace of N which is a union of cells of K. A *subdivision of K* is another cellular decomposition K' of N such that each cell of K' is contained in a cell of K.

Triangulations are cellular decompositions with additional properties. Hence, all previous definitions make sense also for triangulations.

Define the *standard i-simplex* as

$$\Delta^i = \big\{ (t_0, \ldots, t_i) \in \mathbb{R}^{i+1} \mid t_0 + \cdots + t_i = 1 \text{ and } t_k \geq 0 \text{ for all } k \big\}.$$

A regular cellular decomposition T of N is a *triangulation* if the characteristic map $\hat{g}_\kappa$ of any i-cell $\kappa \in T$ is now a homeomorphism $\hat{g}_\kappa : \Delta^i \to \mathrm{cl}(\kappa)$ that sends faces of Δ^i to faces of κ. If T is a triangulation of N, the pair (N, T) is a *simplicial complex*

Remark 1.5. In a triangulation, each i-cell is incident with exactly $i + 1$ $(i - 1)$-cells, and their adjacencies are like those of the faces of the i-simplex.

An i-cell of a triangulation T of N is an *i-simplex* of T. It is convenient to refer to the 0-, 1-, 2- and 3-simplices of T as *vertices*, *edges*, *triangles* and *tetrahedra* of T, respectively.

The subcomplexes of a triangulation T are called *simplicial subsets with respect to T*. If N and N' are manifolds with triangulations T and T' respectively, a map $f : N \to N'$ is *simplicial with respect to T and T'* if for every simplex $\kappa \in T$ with vertices $\{V_1, \ldots, V_i\}$, $f(\kappa)$ is a simplex of T' whose vertices are $\{f(V_1), \ldots, f(V_i)\}$.

Remark 1.6. In our notation, a cell of a cellular decomposition (resp. a simplex of a triangulation) is what is usually called *open cell* (resp. *open simplex*).

By abuse of notation, a cellular or simplicial complex (N, T) is often denoted by N when no confusion can arise about the particular cellular decomposition or triangulation of N that is being considered.

Chapter 2

Filling Dehn surfaces

2.1 Dehn surfaces in 3-manifolds

Let M be a 3-manifold. A subset Σ of M is a *Dehn surface* if there exists a surface S and a normal transverse immersion $f : S \to M$ such that $f(S) = \Sigma$ (see [51]). The map f is called a *parametrization of* Σ.

Since a parametrization f of a Dehn surface is normal and transverse, its singular points must be like those depicted in Fig. 1.3(a) and Fig. 1.3(b) of Sec. 1.4. Moreover, triple points are isolated, and the singular set is a union of curves. By compactness of S, these curves are closed.

Definition 2.1 ([62]). A *double curve* of a Dehn surface $\Sigma \subset M$ is a closed curve in M contained in the singular set of f.

Remark 2.2. Since S is compact, there are only a finite number of double curves in Σ.

There are six branches of double curves which are incident with a triple point. Since closed curves are assumed to be smoothly immersed, when arriving to a triple point P through one of those six branches, the "opposite" branch through which the curve continues is naturally determined (see Fig. 1.3(b)).

The characterization of the singularities of a normal transverse immersion implies that if $g : S' \to M$ is another parametrization of Σ, a triple point of f is also a triple point of g; a double point of f is a double point of g; and a double curve of f is a double curve of g. Moreover, it is straightforward to construct a diffeomorphism $\phi : S \to S'$ such that $f = g \circ \phi$. In other words, the domain surface S, the singular set $\text{Sing}(f)$ of f, the set of triple points of f and the set of double curves of f only depends on the Dehn surface Σ, and not on the chosen parametrization. Therefore, it makes sense to

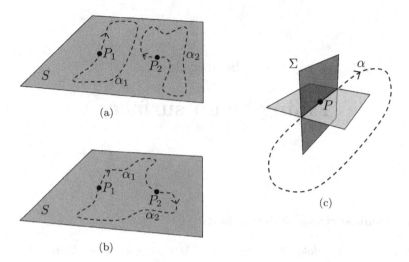

(a)

(b)

(c)

Fig. 2.1: The two possibilities for the lifts of a double curve.

say that the surface S is the *domain of* Σ, and to define *double points* and *triple points* of Σ as the double and triple points of any parametrization of Σ. We also define the *singular set* $\text{Sing}(\Sigma)$ of Σ and the *triple point set* $\text{T}(\Sigma)$ of Σ as the singular set and triple point set of any parametrization of Σ, respectively. Points in $\Sigma - \text{Sing}(\Sigma)$ are called *simple*. If the domain of Σ is a 2-sphere, Σ is called a *Dehn sphere*, if it is a torus, Σ is called a *Dehn torus*, if it is a g-torus, Σ is called a *Dehn g-torus*, and so on. The first studies on Dehn surfaces were on *Dehn disks*, which are 2-disks in 3-manifolds where Dehn's Lemma applies [51]. A particular case of Dehn surfaces are the embedded ones. A Dehn sphere Σ of M is a *standardly embedded 2-sphere* in M if it is the boundary of an embedded 3-disk in M. If the domain surface of Σ is non-connected, a *component* of Σ is the image of a connected components of S under a parametrization of Σ.

Let P be a double point of Σ, and let $P_1, P_2 \in S$ be the preimages of P under the parametrization f. Let $\bar{\alpha}$ be the double curve of Σ passing through P. Making P run along $\bar{\alpha}$, the preimages P_1 and P_2 of P describe two trajectories along pre-$\text{Sing}(f)$. When P returns to its initial position (double curves are always closed), there are two possibilities for the trajectories of P_1 and P_2:

Type I double curve. Both P_1 and P_2 return to their initial positions as in Fig. 2.1(a). Therefore the preimage of the double curve $\bar{\alpha}$ is the union of two closed curves α_1 and α_2 in S, both projected over $\bar{\alpha}$ by f.

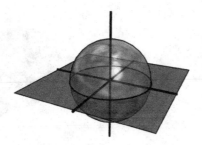

Fig. 2.2: The standardly embedded $\mathbb{S}^2$ in $\mathbb{S}^3$.

Type II double curve. The points P_1 and P_2 swap their positions as in Fig. 2.1(b). In this case, $f^{-1}(\bar{\alpha})$ is a single curve in S that doubly covers $\bar{\alpha}$.

This notation was introduced by Johansson [36].

Lemma 2.3 ([36]). *If S is orientable, type II double curves are orientation-reversing in M.* □

Therefore, if both S and M are orientable there are no type II double curves. We restrict our analysis to this case, so all double curves can be safely assumed to be of type I.

Definition 2.4. Let $\bar{\alpha} : \mathbb{S}^1 \to M$ be a parametrization of a double curve of Σ. A *lift of $\bar{\alpha}$ to S by f* is a parametrized closed curve $\alpha : \mathbb{S}^1 \to S$ such that $f \circ \alpha = \bar{\alpha}$. Two distinct lifts of the same double curve $\bar{\alpha}$ are called *sister (parametrized) curves*. Two closed curves in pre-Sing(f) are *sister curves*[1] if they have sister parametrizations.

Remark 2.5. Since all double curves are assumed to be of type I, each of them admits exactly two lifts.

2.2 Filling Dehn surfaces

Definition 2.6 ([49]). A Dehn surface Σ in a 3-manifold M *fills* M if it defines a cellular decomposition of M for which:

(1) $T(\Sigma)$ is the 0-skeleton;
(2) Sing(Σ) is the 1-skeleton; and
(3) Σ is the 2-skeleton.

---

[1]Sister curves are the *companion* curves of [11].

(a)   (b)

Fig. 2.3: Immersed surfaces in $\mathbb{S}^3$. (a) Non-filling immersion. (b) Filling immersion.

In this case we also say that Σ is *filling* in M. If f is a parametrization of a filling Dehn surface Σ, we also refer to f as a *filling* immersion, and we say that f *fills* M.

The following proposition gives an equivalent definition of filling Dehn surface which will be useful later. Its proof is straightforward.

Proposition 2.7. *The Dehn surface Σ fills M if and only if:*

(1) $M - \Sigma$ is a disjoint union of open 3-balls;
(2) $\Sigma - \mathrm{Sing}(\Sigma)$ is a disjoint union of open disks; and
(3) $\mathrm{Sing}(\Sigma) - \mathrm{T}(\Sigma)$ is a disjoint union of open arcs. $\square$

Example 2.8. A 2-sphere Σ standardly embedded in $\mathbb{S}^3$ is not filling: even when condition 1 of Proposition 2.7 is true, it does not satisfy conditions 2 and 3 since both $\mathrm{Sing}(\Sigma)$ and $\mathrm{T}(\Sigma)$ are empty.

Example 2.9. A couple of embedded 2-spheres in $\mathbb{S}^3$ meeting along a single double curve as in Fig. 2.3(a) is not filling because its triple point set is empty. In this case, conditions 1 and 2 of Proposition 2.7 are satisfied.

Example 2.10. Let Σ be the union of three embedded 2-spheres in $\mathbb{S}^3$ as depicted in Fig. 2.3(b). The Dehn surface Σ fills $\mathbb{S}^3$.

As it can be seen in Fig. 2.3(b), each pair of spheres are like those in Fig. 2.3(a), but the three meet at two triple points. The three conditions of the Proposition 2.7 are fulfilled. This seems to be the simplest example of filling Dehn surface.

Haken proved that all homotopy 3-sphere M has a filling Dehn sphere [23], and that it can be constructed deforming a embedded 2-sphere in M (see [24]).

Later, a weaker version of this statement but valid for any manifold was proved in [16]: any 3-manifold has a Dehn sphere Σ such that $M - \Sigma$ is a disjoint union of open 3-balls, that is, fulfilling condition 1 of Proposition 2.7. Finally, Montesinos extended Haken's result for any 3-manifold:

Theorem 2.11 ([49]). *Any 3-manifold has a null-homotopic filling Dehn sphere.*

A Dehn surface is *null-homotopic* if it has a null-homotopic parametrization. As it is pointed out in [69] the filling Dehn spheres constructed in [49] are null-homotopic. Simpler proofs of this result can be seen in [3, 67]. We recommend the reader to check the short and extremely elegant proof given in [3].

A weaker version of filling Dehn surfaces are the *quasi-filling Dehn surfaces* defined in [4], which are Dehn surfaces whose complementary set in M is a disjoint union of open 3-balls. The Dehn surfaces of Examples 2.8 and 2.9 are quasi-filling in $\mathbb{S}^3$. With this notation, the result in [16] is that every 3-manifold has a quasi-filling Dehn sphere.

2.3 Notation

If Σ is a filling Dehn surface of M, a *cell* (resp. *i-cell*) of Σ is a cell (resp. *i*-cell) of the cellular decomposition of M defined by Σ. We will also use the following notation:

- an *edge*[2] of Σ is a connected component of $\mathrm{Sing}(\Sigma) - \mathrm{T}(\Sigma)$
- a *face* of Σ is a connected component of $\Sigma - \mathrm{Sing}(\Sigma)$; and
- a *region* of Σ is a connected component of $M - \Sigma$.

Hence, the edges, faces and regions of Σ are the 1-, 2- and 3-cells of the cellular decomposition of M defined by Σ. The 0-cells of that cellular decomposition are the triple points of Σ. The adjacency and incidence relations introduced in Sec. 1.6 are naturally translated to the present notation. By an abuse of notation, we sometimes use the terms "region", "face" and "edge" of Σ as defined here even when Σ is not filling.

Additionally, a filling Dehn surface Σ is *regular* if the cellular decomposition of M induced by Σ is regular.

---

[2]This notion does not agree with the notion of edge of a triangulation, because in this case the closure of the edge can be a circle. While triangulations are always assumed to be regular, cell decompositions defined by filling Dehn surfaces are not necessarily regular.

Let κ and κ' be distinct cells of Σ with $\kappa < \kappa'$. The cell κ' is *regular at* κ if for every point $Q \in \kappa$ there exists a neighbourhood of Q whose intersection with κ' is connected. Otherwise, κ' is *self-adjacent in* κ. It is easy to see that κ is regular in the sense of Sec. 1.6 if and only if it is regular at all its faces, which is equivalent to being regular only at its vertices.

Hence, in order to check the regularity of a filling Dehn surface Σ it is enough to see what happens around its triple points. If P is a triple point of Σ, the surface Σ is *regular at* P if P is incident with 6 distinct edges, 12 different faces and 8 separate regions of Σ. Then, the filling Dehn surface Σ is regular if it is regular at all its triple points.

If κ is a regular i-cell of Σ with $i > 1$, the closure $\mathrm{cl}(\kappa)$, which is homeomorphic to $\mathbb{D}^i$, has an induced structure of manifold *with boundary with corners* embedded in M [13,46]. This "cornered structure" makes $\mathrm{cl}(\kappa)$ look like a polygon or polyhedron of classical geometry, so we will use a similar notation to that of polygons and polyhedra. If there are n edges of Σ incident with κ, the set $\mathrm{cl}(\kappa)$ is an n-gon of Σ and κ is an *open n-gon*. In the context of filling Dehn surfaces, there exist n-gons for any $n \geq 1$. For example, all the faces of the filling Dehn surface of $\mathbb{S}^3$ depicted in Fig. 2.3(b) are 2-gons.

If κ is a regular region of Σ, there are many possible configurations for $\mathrm{cl}(\kappa)$. The cells of Σ contained in $\partial\kappa$ form a cellular decomposition of the 2-sphere $\partial\kappa$. Since there are three edges contained in $\partial\kappa$ incident in each triple point in $\partial\kappa$, the 1-skeleton of this cellular decomposition of $\partial\kappa$ is a trivalent graph. Some different configurations of $\mathrm{cl}(\kappa)$ that may occur are:

n-gonal prism. The set $\mathrm{cl}(\kappa)$ is diffeomorphic to the product $w \times I$ where w is a n-gon for some $n \geq 1$ (see Figs 2.4(a), (b) and (c)).

Tetrahedron. The set $\mathrm{cl}(\kappa)$ is diffeomorphic to a rectilinear tetrahedron in the euclidean space.

Trihedron. $\mathrm{cl}(\kappa)$ has three 2-gonal faces (see Fig. 2.4(d)).

As in the case of n-gons, when we have faces that do not exist in classical "rectilinear" geometry, like 1-gons or 2-gons, there appear some "polyhedra" with no counterpart in affine geometry. Examples of these "essentially curvilinear polyhedra" are the 1- and 2-gonal prisms depicted in Figs 2.4(a) and (b) and the trihedron of Fig. 2.4(d). Each region of the filling Dehn surface of Fig. 2.3(b) is a trihedron. The region κ will be an open n-gonal prism, an open tetrahedron, an open trihedron, etcetera, if $\mathrm{cl}(\kappa)$ is an n-gonal prism, a tetrahedron, a trihedron, etcetera, respectively.

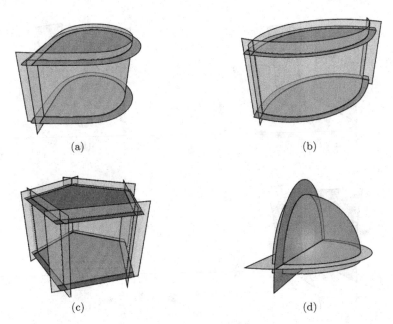

(a)

Fig. 2.4: Common prisms found as regions of filling Dehn surfaces.

2.4 Surgery on Dehn surfaces. Montesinos Theorem

Banchoff introduces in [7] the modifications of Dehn surfaces by *surgery*, which are just adaptation to Dehn surfaces of the *piping* operations [58].

Surgery on Dehn surfaces is like a connected sum adapted to the singularities of the surface: two disjoint small open 2-disks are removed and the resulting circles are glued together with a tube. This modification can be done in quite general situations, but we will cover only the cases introduced in [7], where both the disks and the tube are in a very precise position with respect to the rest of the Dehn surface.

From now until the end of the chapter Σ denotes a Dehn surface in a manifold M and $f : S \to M$ denotes a parametrization of Σ. As usual, $\mathbb{D}^2$ is the standard closed unit 2-disk in $\mathbb{R}^2$, $I = [0, 1]$ is the closed unit interval and $\partial \mathbb{D}^2 = \mathbb{S}^1$. Let $\mathbf{O}$ denote the centre of $\mathbb{D}^2$, and I_x and I_y denote the intersection of $\mathbb{D}^2$ with the horizontal and vertical coordinate axes of $\mathbb{R}^2$, respectively.

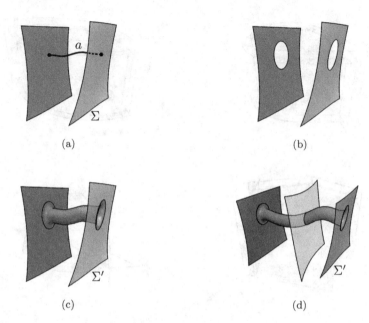

(a)   (b)

(c)   (d)

Fig. 2.5: Type 0 surgery along an arc.

2.4.1 *Type 0 arcs*

Definition 2.12. An arc a in M is of *type 0 with respect to* Σ (Fig. 2.5) if

(1) $a \cap \Sigma$ is a finite set containing the endpoints of a;
(2) $a \cap \mathrm{Sing}(\Sigma) = \emptyset$; and
(3) a is transverse to Σ at every point of $a \cap \Sigma$.

Moreover, a is *simple* if $a \cap \Sigma$ only contains its endpoints.

If a is a type 0 arc with respect to Σ, the set $f^{-1}(a)$ is a finite set disjoint from pre-$\mathrm{Sing}(f)$. Let $Y_0, \ldots, Y_n$ be the points of $a \cap \Sigma$, so ordered that Y_0 and Y_n are the endpoints of a and we successively find $Y_1, \ldots, Y_{n-1}$ when travelling along a from Y_0 to Y_n. Let $\{X_i\}_{i=0}^n$ be the points of S such that $f(X_i) = Y_i$ for $i = 0, \ldots, n$. Consider two closed disks D_0 and D_n such that $X_0 \in \mathrm{int}(D_0)$ and $X_n \in \mathrm{int}(D_n)$, $D_i \cap \mathrm{pre\text{-}Sing}(f) = \emptyset$ for $i = 0$ and n, and none of the points $X_1, \ldots, X_{n-1}$ meet the disks D_0 and D_n.

It is possible to find an embedding $\xi : \mathbb{D}^2 \times I \to M$ such that:

(1) the image under ξ of the axis $\{\mathbf{O}\} \times I$ of the cylinder $\mathbb{D}^2 \times I$ is a;

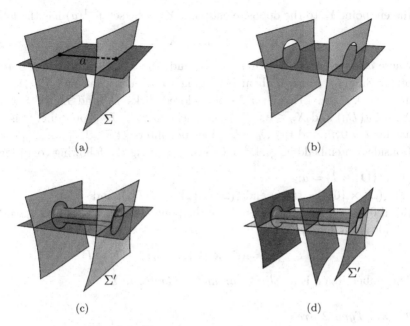

(a)

(b)

(c)

(d)

Fig. 2.6: Type 1 surgery along an arc.

(2) $\xi(\mathbb{D}^2 \times \{0\}) = f(D_0)$ and $\xi(\mathbb{D}^2 \times \{1\}) = f(D_n)$; and
(3) If $0 = t_0 < t_1 < \cdots < t_n = 1$ are such that $\xi(\mathbf{0}, t_i) = Y_i$ for all $i = 0, \ldots, n$, then $\xi^{-1}(\Sigma) = \mathbb{D}^2 \times \{t_0, \ldots, t_n\}$.

The embedding ξ is a *cylinder around a adapted to Σ*.

2.4.2 Type 1 arcs

Definition 2.13. The arc a in M is of *type 1 with respect to Σ* if

(1) $a \subset \Sigma$;
(2) $a \cap \mathrm{Sing}(\Sigma)$ is a finite set containing the endpoints of a;
(3) $a \cap \mathrm{T}(\Sigma) = \emptyset$; and
(4) a is nowhere tangent to $\mathrm{Sing}(\Sigma)$.

The arc a is *simple* if $a \cap \mathrm{Sing}(\Sigma)$ only contains its endpoints.

Let a be a type 1 arc in M with respect to Σ, and let $Y_0, \ldots, Y_n$ be the points of $a \cap \mathrm{Sing}(\Sigma)$ ordered, as they are found when traveling along a from the endpoint Y_0 to the opposite endpoint Y_n. The set $f^{-1}(a)$ has the form

$$\widetilde{a} \cup \{X_0, \ldots, X_n\},$$

where $\widetilde{a}$ is an arc in S with $f(\widetilde{a}) = a$ and $X_0, \ldots, X_n$ are distinct points of pre-Sing(f), none of them belonging to $\widetilde{a}$, such that $f(X_i) = Y_i$ for all $i = 0, \ldots, n$. Take two disjoint closed disks D_0 and D_n of S with $X_0 \in \text{int}(D_0)$ and $X_n \in \text{int}(D_n)$ and such that: (i) $D_i \cap \text{pre-Sing}(f)$ is an arc for $i = 0, n$; and (ii) $D_0 \cup D_n$ does not intersect $\widetilde{a} \cup \{X_1, \ldots, X_n - 1\}$. Consider an embedding $\xi : \mathbb{D}^2 \times I \to M$ satisfying the following conditions:

(1) $\xi(\{\mathbf{O}\} \times I) = a$;
(2) $\xi(\mathbb{D}^2 \times \{0\}) = f(D_0)$ and $\xi(\mathbb{D}^2 \times \{1\}) = f(D_n)$; and
(3) If $0 = t_0 < t_1 < \cdots < t_n = 1$ are the points of I such that $\xi(\mathbf{O}, t_i) = Y_i$ for $i = 0, 1, \ldots, n$, then

$$\xi^{-1}(\Sigma) = (\mathbb{D}^2 \times \{t_0, \ldots, t_n\}) \cup (I_x \times I).$$

The embedding ξ is a *cylinder around a adapted to* Σ.

2.4.3 Type 2 arcs

Definition 2.14. An arc a in M is of *type 2 with respect to* Σ if

(1) $a \subset \text{Sing}(\Sigma)$; and
(2) the endpoints of a are triple points of Σ.

The arc a is *simple* if $a \cap \text{T}(\Sigma)$ only contains its endpoints.

Let a be a type 2 arc in M with respect to Σ. Denote by $Y_0, \ldots, Y_n$ the points of $a \cap \text{T}(\Sigma)$, ordered as in the previous two cases. Now,

$$f^{-1}(a) = \widetilde{a}_1 \cup \widetilde{a}_2 \cup \{X_0, \ldots, X_n\},$$

where $\widetilde{a}_1$ and $\widetilde{a}_2$ are two disjoint arcs in pre-Sing(f) such that $f(\widetilde{a}_i) = a$ for $i = 1, 2$, and $X_0, \ldots, X_n$ are distinct crossings of pre-Sing(f) disjoint from $\widetilde{a}_1 \cup \widetilde{a}_2$ such that $f(X_i) = Y_i$ for $i = 0, \ldots, n$.

As before, take two disjoint closed disks D_0, D_n of S with $X_0 \in \text{int}(D_0)$ and $X_n \in \text{int}(D_n)$ such that

$$(D_0 \cup D_n) \cap (\widetilde{a}_1 \cup \widetilde{a}_2 \cup \{X_1, \ldots, X_{n-1}\}) = \emptyset$$

and such that for $i = 0, n$ the 2-disk D_i intersects pre-Sing(f) in a "cross" built from two arcs intersecting in X_i.

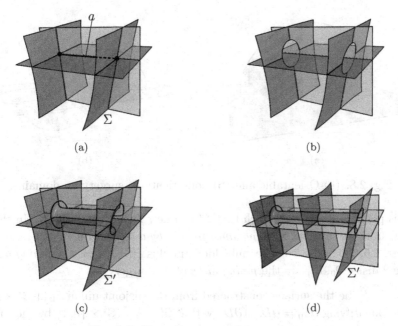

$$(a)$$ $$(b)$$

$$(c)$$ $$(d)$$

Fig. 2.7: Type 2 surgery along an arc.

Consider an embedding $\xi : \mathbb{D}^2 \times I \to M$ satisfying:

(1) $\xi(\{\mathbf{O}\} \times I) = a$;

(2) $\xi(\mathbb{D}^2 \times \{0\}) = f(D_0)$ and $\xi(\mathbb{D}^2 \times \{1\}) = f(D_n)$: and

(3) if $0 = t_0 < t_1 < \cdots < t_n = 1$ are the points in I such that $\xi(\mathbf{O}, t_i) = Y_i$
for $i = 0, \ldots, n$, then

$$\xi^{-1}(\Sigma) = \big(\mathbb{D}^2 \times \{t_1, \ldots, t_n\}\big) \cup \big((I_x \cup I_y) \times I\big).$$

Again, ξ is a *cylinder around a adapted to* Σ.

2.4.4 *Surgeries*

Let a by a type i arc in M with respect to Σ, for some $i = 0, 1, 2$. Consider
the 2-disks $D_0, D_n \subset S$ and a cylinder $\xi : \mathbb{D}^2 \times I \to M$ around a adapted
to Σ as introduced above.

Let S_0 be the surface with boundary that is obtained by removing the
interior of D_0 and D_n from S, and define the surface

$$\Sigma' = f(S_0) \cup \xi(\mathbb{S}^1 \times I).$$

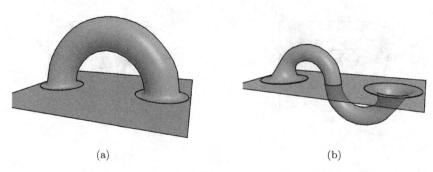

(a)   (b)

Fig. 2.8: (a) Orientable and (b) non-orientable surgery on domains.

It is possible to choose ξ such that Σ' is a new Dehn surface in M. In this situation, we say that Σ' *is obtained from* Σ *by a type i surgery along a* (see Figs. 2.5, 2.6 and 2.7). The embedded annulus $\xi(\mathbb{S}^1 \times I)$ is the *piping* and the 2-disks D_0, D_n are the *piping disks* of the surgery.

Let S' be the surface constructed from the disjoint union $S_0 \sqcup (\mathbb{S}^1 \times I)$ after identifying $\partial S_0 = \partial D_0 \cup \partial D_n$ with $\partial(\mathbb{S}^1 \times I) = \mathbb{S}^1 \times \{0, 1\}$ by the rule

$$x \in \partial S_0 \simeq y \in \partial \mathbb{S}^1 \times I \iff f(x) = \xi(y).$$

Let $f' : S' \to M$ be the map that agrees with f on S_0 and with ξ on $\mathbb{S}^1 \times I$. By the definition of $\simeq$, f' is well-defined, and we say also that f' *is obtained from* f *by a type i surgery along a*.

If D_0 and D_n lie in the same connected component of S, then S' is the surface resulting from S after adding a handle to this connected component. If, as usual, S is assumed to be orientable, *the surface S' might be non-orientable*. Thus, to ensure orientability of domains, the surgery should be carefully done. Figure 2.8 shows a simple example of this situation: the surgery of Fig. 2.8(a) preserves orientability while the surgery depicted in Fig. 2.8(b) destroys orientability.

However, if D_0 and D_n are in different connected components of S, then S' is the result of performing the connected sum of these connected components in S. In this case, if S is orientable, S' is also orientable and it has fewer connected components than S. Moreover, if S *is a disjoint union of k 2-spheres, S' is a disjoint union of $(k-1)$ 2-spheres*.

We will frequently use that property: starting with a filling Dehn surface whose domain is a union of spheres, it is possible to build, by surgery, a filling Dehn sphere in M.

Proposition 2.15. *Let* Σ *and* Σ' *be two Dehn surfaces of* M *and assume that* Σ *fills* M*. Then*

(1) If Σ' *is obtained from* Σ *by a type 0 surgery, then* Σ' *does not fill* M*;*

(2) If Σ' *is obtained from* Σ *by a simple surgery of type 1 or type 2, then* Σ' *might or might not fill* M*; and*

(3) If Σ' *is obtained from* Σ *by a non simple surgery of type 1 or 2, then* Σ' *fills* M*.*

Proof. Let a be the type i arc in M with respect to Σ along which the surgery that transforms Σ in Σ' is performed.

(1) If a is of type 0 and R is a region of Σ meeting a, let R' be the connected component of $M - \Sigma'$ that agrees with R except around a. Since Σ fills M, the region R is homeomorphic to an open 3-ball. For each connected component of $a \cap R$, we "drill" R removing an open cylinder from it, so R' cannot be an open 3-ball.

(2) If a is of type 1 or 2 and simple, the piping could connect with itself a region or a face of Σ. If this is the case, the surface Σ' no longer fills M.

(3) If a is not simple and it is of type 1 or 2, the self-connections on faces or regions of Σ mentioned in (2) are prevented by the faces of Σ transverse to a at the interior points of a. By Proposition 2.7, in this case Σ' is always a filling Dehn surface. $\square$

2.4.5 *Spiral piping*

Proposition 2.15 gives a way to modify filling Dehn surfaces preserving fillingness. A particular interesting surgery is the *spiral piping* around a triple point. If P is a triple point of Σ, it is possible to define another Dehn surface Σ' modifying Σ in a sufficiently small neighbourhood of P as in Fig. 2.9. In this case, we say that Σ' is obtained from Σ by a *spiral piping around P*. This modification is in fact a non simple type 1 surgery, and hence if Σ fills M, Σ' will fill M too.

Spiral pipings have very good properties that make them extremely useful. They are used throughout the book, but specially in the constructions of Chap. 6.

Proposition 2.16. *Let* Σ *be a filling Dehn surface of* M *and let* Σ' *be obtained from* Σ *by a spiral piping. If* Σ *is regular, then* Σ' *is also regular.*

Proof. Let P be the triple point of Σ around which the spiral piping has been carried out to obtain Σ'. We can assume without loss of generality

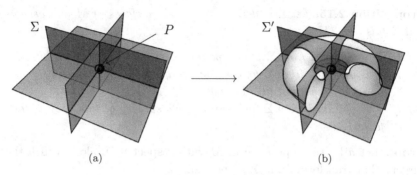

(a)   (b)

Fig. 2.9: Spiral surgery.

that P and the surgery around it are like those in Fig. 2.9. The cells of Σ not meeting P are cells of Σ' without modifications.

Since Σ is regular, it is regular at P. There are 8 different regions, 12 faces and 6 edges meeting at P. It is clear from Fig. 2.9 that the surgery creates no self-adjacencies in the cells of Σ' coming from those of Σ meeting at P.

Finally, the "new" cells of Σ' around P generated by the surgery are all regular. So, all the cells of Σ' are regular. □

Definition 2.17 ([67]). A filling Dehn surface of a 3-manifold is a *filling collection of spheres* if its domain is a disjoint union of 2-spheres.

Theorem 2.18 ([67]). *If a 3-manifold has a filling collection of spheres, it has a filling Dehn sphere.*

Proof. Let Σ be a filling collection of spheres in a 3-manifold M. By Def. 2.17, the domain of Σ is a disjoint union of 2-spheres $S = \bigsqcup_{i=1}^{m} \mathbb{S}_i^2$, for some $m \in \mathbb{N}$. Let $f : S \to \Sigma \subset M$ be a parametrization of Σ.

If $m = 1$ there is nothing to prove. Assume $m > 1$. The filling Dehn sphere Σ is the union of the Dehn spheres $\Sigma_i = f(\mathbb{S}_i^2)$ with $i = 1, \dots, m$.

Since M is connected, the 2-skeleton of any cellular decomposition of M is also connected, therefore Σ is connected. This implies that there exists two indices $i \neq j$ such that $\Sigma_i \cap \Sigma_j \neq \emptyset$. Since Σ is a Dehn sphere, Σ_i and Σ_j meet, at least, along a double curve a of Σ, and since Σ is filling, a contains a triple point P of Σ. In this triple point P, the two Dehn spheres Σ_i and Σ_j intersect a third Dehn sphere Σ_k with $k \in \{1, \dots, m\}$ (k may be equal to i or j). It is possible to connect Σ_i with Σ_j by a spiral piping around P. The Dehn surface obtained from Σ after this operation is a filling collection of spheres with $m - 1$ spheres in its domain. Repeating the process $m - 2$ times we obtain a filling Dehn sphere of M. □

Therefore, to proof Theorem 2.11 it is enough to build a filling collection of spheres in any 3-manifold M. This is indeed the technique used in [49, 67], but without using the spiral piping to connect the spheres. In both works a filling collection of spheres is constructed starting from a Heegaard decomposition of the 3-manifold. In Sec. 6.4, a filling collection of spheres will be constructed from any triangulation of a 3-manifold, which gives a another proof of Theorem 2.11.

Chapter 3

Johansson diagrams

3.1 Diagrams associated to Dehn surfaces

Let Σ be a Dehn surface in M and consider a parametrization $f : S \to M$ of Σ.

Definition 3.1 (Johansson diagram 1 [49]). The pre-singular set of f with the information of how its points are identified by f is called the *Johansson diagram* of Σ.

Although this definition depends on f, different parametrizations produce diagrams which are equivalent up to a diffeomorphism of S. The Johansson diagram of Σ naturally defines an equivalence relation "$\sim$" on S:

two points of S are related by $\sim$ if they are related by f.

The Dehn surface Σ, as a topological space, is homeomorphic to $S/\sim$. Therefore, the Johansson diagram allows to study some intrinsic properties of Σ independently of the immersion f and the manifold M.

Definition 3.1 is a bit ambiguous, since the gluing information on the set pre-Sing(f) can be given in multiple ways. We will use the double curves of Σ to give a more precise definition.

A *parametrization* of the singular set Sing(Σ) is a set of parametrized curves $\bar{\mathcal{D}} = \{\bar{\alpha}_1, \bar{\alpha}_2, \ldots, \bar{\alpha}_m\}$ such that

(1) $\bar{\alpha}_i$ parametrizes a double curve of Σ for all $i = 1, \ldots, m$;
(2) Sing(Σ) $= |\bar{\mathcal{D}}|$; and
(3) $|\bar{\alpha}_i| \neq |\bar{\alpha}_j|$ if $i \neq j$.

Let $\bar{\mathcal{D}}$ be a parametrization of Sing(Σ) and let $\mathcal{D}$ be the collection of all lifts to S of the curves of $\bar{\mathcal{D}}$. Since f is transverse, $\mathcal{D}$ is a normal system

25

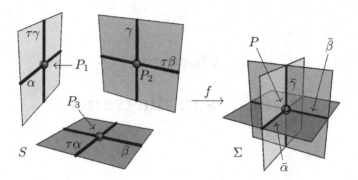

Fig. 3.1: The standard triple point configuration.

of curves in S. The map $\tau : \mathcal{D} \to \mathcal{D}$ that assigns to each curve of $\mathcal{D}$ its sister curve under f is a free involution, that is, $\tau^2\alpha = \alpha$ and $\tau\alpha \neq \alpha$ for all $\alpha \in \mathcal{D}$. The pair $(\mathcal{D}, \tau)$ contains all the information about the singular set of f and how its points are related by f. By construction pre-Sing$(\Sigma) = |\mathcal{D}|$ and two points P and $Q \in S$ are related by f if and only if there exists a curve $\alpha \in \mathcal{D}$ and $t \in \mathbb{S}^1$ such that $P = \alpha(t)$ and $Q = \tau\alpha(t)$. The images of the curves of $\mathcal{D}$ only intersect, transversely, at the crossings of $\mathcal{D}$, which are the preimages under f of the triple points of Σ.

Let P be a triple point of Σ. Three double curves $\bar{\alpha}$, $\bar{\beta}$ and $\bar{\gamma}$ of Σ meet at P. The preimage of P under f is composed by three crossings P_1, P_2 and P_3 of $\mathcal{D}$. The set $\{P_1, P_2, P_3\}$ is called the *triplet* of P in $\mathcal{D}$. Let α, $\tau\alpha$, β, $\tau\beta$, γ and $\tau\gamma$ be the lifts through f of the curves $\bar{\alpha}$, $\bar{\beta}$ and $\bar{\gamma}$ respectively. Renaming these curves if necessary, we can assume that these curves meet around P_1, P_2 and P_3 as in Fig. 3.1:

P_1 is the intersection of α and $\tau\gamma$,

P_2 is the intersection of γ and $\tau\beta$,

P_3 is the intersection of β and $\tau\alpha$.

This combinatorial scheme of $\mathcal{D}$ around the triplet of a triple point of Σ is called the *standard triple point configuration*.

3.2   Abstract diagrams on surfaces

The Johansson Theorem [37] (Theorem 3.15 below) gives necessary and sufficient conditions for a pair $(\mathcal{D}, \tau)$ as above to be the Johansson diagram of a Dehn surface in a 3-manifold. While in [37] only Dehn disks are considered, the proof works for general Dehn surfaces in 3-manifolds if both surfaces and 3-manifolds are orientable.

Definition 3.2 (Abstract diagram, cf. [51]). Let S be a surface, $\mathcal{D}$ a normal system of curves on S and τ a free involution of $\mathcal{D}$. The points P and $Q \in S$ are *related* by $(\mathcal{D}, \tau)$, denoted by $P \sim_{\mathcal{D}} Q$, if $P = Q$ or if there exists $t \in \mathbb{S}^1$ and $\alpha \in \mathcal{D}$ such that $P = \alpha(t)$ and $Q = \tau\alpha(t)$.

The pair $(\mathcal{D}, \tau)$ is a *diagram in* S if

(D1) crossings are always related with crossings; and
(D2) crossings are grouped in *triplets*, sets of three crossings whose elements are pairwise related by $\mathcal{D}$.

Let $(\mathcal{D}, \tau)$ be a diagram in S. Since τ is an involution, $\sim_{\mathcal{D}}$ is symmetric, and (D2) implies that it is reflexive too. Hence, if $(\mathcal{D}, \tau)$ is a diagram in S, $\sim_{\mathcal{D}}$ is an equivalence relation in S. We say that τ is the *sistering* of $(\mathcal{D}, \tau)$ and that α and $\tau\alpha$ are *sister curves* of the diagram $(\mathcal{D}, \tau)$ for each $\alpha \in \mathcal{D}$. If the pair $(\mathcal{D}, \tau)$ is the Johansson diagram of a Dehn surface in a 3-manifold, it fulfills conditions (D1) and (D2). In particular, (D2) is a consequence of the standard triple point configuration.

Example 3.3. Figure 3.2 shows few simple examples of diagrams. Figures 3.2(a) and 3.2(c) are the diagrams that appeared in the seminal works of Johansson [36, 37]. Figure 3.2(b) is a slight modification of the first one. Figures 3.2(e) and 3.2(f) show two distinct diagrams with the same underlying graph. All of them are diagrams on the 2-sphere $\mathbb{S}^2$, except for the last two (Fig. 3.2(g) and 3.2(h)) which are diagrams on the torus $\mathbb{T}^2$. Diagram in Fig. 3.2(h) appears in [69].

In those diagrams, crossings belonging to the same triplet are labelled with the same letter and segments marked with similar arrows are identified following the direction indicated by them.

Definition 3.4. A diagram $(\mathcal{D}, \tau)$ in a surface S is *realizable* if there exists a 3-manifold M and an immersion $f : S \to M$ such that two points P and $Q \in S$ are related by $(\mathcal{D}, \tau)$ if and only if they are related by f. In this situation, we say that f *realizes* $(\mathcal{D}, \tau)$.

Remark 3.5. Given a pair $(\mathcal{D}, \tau)$, where $\mathcal{D}$ is a normal system of curves in a surface S and τ a free involution of $\mathcal{D}$, it is possible to define a topological space $\widetilde{\Sigma}$ by identifying the points of S related by $(\mathcal{D}, \tau)$ as indicated in Definition 3.2. If $(\mathcal{D}, \tau)$ is a diagram, $\widetilde{\Sigma} = S/\sim_{\mathcal{D}}$ is quite similar to a Dehn

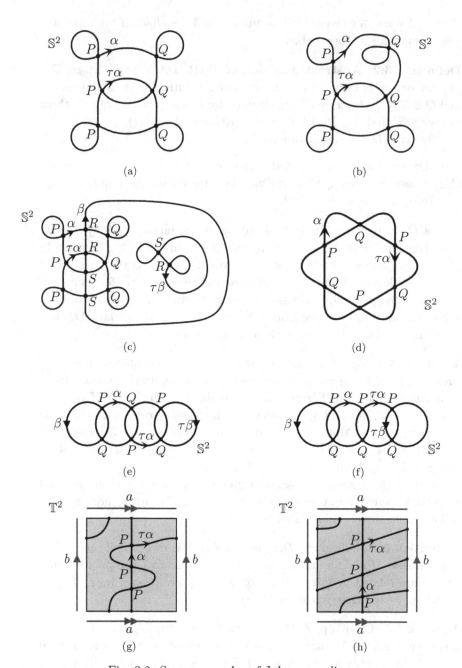

Fig. 3.2: Some examples of Johansson diagrams.

surface: it has double and triple points, double curves, etcetera, and the diagram $(\mathcal{D}, \tau)$ will be realizable if and only if $\widetilde{\Sigma}$ can be "nicely embedded" in a 3-manifold. The spaces $\widetilde{\Sigma}$ obtained in this way are called *pseudo Dehn surfaces*. Hence, there could be realizable and non-realizable pseudo Dehn surfaces. As it will be seen in Chap. 4, some properties of Dehn surfaces also are valid for pseudo Dehn surfaces.

From now on the diagram $(\mathcal{D}, \tau)$ will be simply denoted by $\mathcal{D}$ leaving the sistering τ implicitly defined.

The notion of abstract diagram allows us to give a new (and more technical) definition of Johansson diagram:

Definition 3.6 (Johansson diagram 2). Let Σ be a Dehn surface in M. The Johansson diagram of Σ is any diagram $\mathcal{D}$ in the domain of Σ realizable by a parametrization of Σ.

A diagram $\mathcal{D}$ in a surface S is *connected* if $|\mathcal{D}|$ is connected. The diagram $\mathcal{D}$ *fills* S if it defines a cellular decomposition of S in which the 0-skeleton is the set of crossings of $\mathcal{D}$ and the 1-skeleton is $|\mathcal{D}|$. Hence, if S is connected and $\mathcal{D}$ fills S, then $\mathcal{D}$ is connected.

Definition 3.7. The diagram $\mathcal{D}$ is *filling* if there exists a 3-manifold M and a filling immersion $f : S \to M$ realizing $\mathcal{D}$.

Proposition 3.8. *If $\mathcal{D}$ is filling, then $\mathcal{D}$ fills S.*

Proof. Let $f : S \to M$ be a filling immersion that realizes $\mathcal{D}$. The immersion f diffeomorphically maps each connected component of $S - |\mathcal{D}|$ onto a connected component of $f(S) - \text{Sing}(f)$. Since f is filling all these connected components should be homeomorphic to disks. Moreover, each double curve of f meets a triple point of f, so each curve of $\mathcal{D}$ must meet a crossing of $\mathcal{D}$. Therefore, the set

$$|\mathcal{D}| - \{\text{crossings of } \mathcal{D}\}$$

is a disjoint union of open arcs. $\qquad\square$

Example 3.9. The diagram of Fig. 3.2(c) cannot be filling because there is a connected component of $\mathbb{S}^2 - |\mathcal{D}|$ which is an annulus.

The proof of the following lemma is straightforward:

Lemma 3.10. *A diagram in $\mathbb{S}^2$ is connected if and only if it fills $\mathbb{S}^2$.* $\qquad\square$

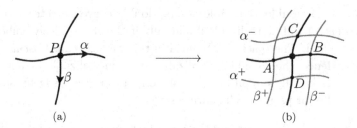

(a)   (b)

Fig. 3.3: Crossing configuration after adding the neighbouring curves.

3.3   The Johansson Theorem

Let S be a connected orientable surface. Consider a diagram $\mathcal{D}$ on S.

Let α be a curve of $\mathcal{D}$. Since S is orientable, the immersion α is 2-sided. Then, there exist two parametrized closed curves, α^- and α^+, running parallel to α at different sides of α. Let assume that both $\mathbb{S}^1$ and S are oriented, and orient α following the positive orientation of $\mathbb{S}^1$. Finally, orient α^- and α^+ in parallel to α.

Definition 3.11. The curve α^- is *on the left hand side* and α^+ is *on the right hand side* of α if for every point $P \in |\alpha|$ which neither is a crossing point of $\mathcal{D}$ nor belongs to $|\alpha| \cap (|\alpha^-| \cup |\alpha^+|)$, there exists an open neighbourhood U_P of P in S and an orientation-preserving diffeomorphism $u_P : U_P \to \mathbb{R}^2$ such that:

(1) $|\alpha| \cap U_P$, with the positive orientation, is mapped onto the x axis, with the positive orientation.
(2) $|\alpha^-| \cap U_P$, with the positive orientation, is mapped onto the line $y = 1$, oriented from left to right.
(3) $|\alpha^+| \cap U_P$, with the positive orientation, is mapped onto the line $y = -1$, oriented from left to right.

We will always assume that α^- is on the left hand side and α^+ is on the right hand side of α.

The curves α^- and α^+ are *opposite* to each other and they are called *neighbouring curves* of α in $\mathcal{D}$.

To construct a set $\mathcal{N}(\mathcal{D})$ containing a pair of neighbouring curves for all the curves of $\mathcal{D}$, we choose the curves of $\mathcal{N}(\mathcal{D})$ such that $\mathcal{D} \cup \mathcal{N}(\mathcal{D})$ is a normal system of curves in S whose crossings only appear near the crossings of $\mathcal{D}$, and following the configuration of Fig. 3.3. As it can be seen in that figure, if $\mathcal{D}$ has d crossings, $\mathcal{D} \cup \mathcal{N}(\mathcal{D})$ has exactly $9d$ crossings.

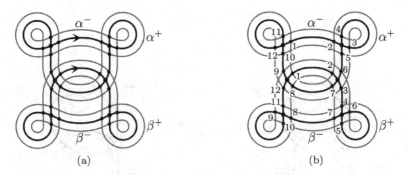

Fig. 3.4: Neighbouring points of a diagram.

The points of $|\mathcal{D}| \cap |\mathcal{N}(\mathcal{D})|$ are called *neighbouring points of* $\mathcal{D}$. Around a crossing P of $\mathcal{D}$ there are four neighbouring points of $\mathcal{D}$ (the points A, B, C, D in Fig. 3.3(b)) which are called *neighbouring points of* P. The curves of $\mathcal{N}(\mathcal{D})$ should satisfy a last property similar to the first condition of Definition 3.2:

Neighbouring points of $\mathcal{D}$ *are related by* $\mathcal{D}$ *with neighbouring points of* $\mathcal{D}$.

Example 3.12. Consider the original diagram of Johansson [36] depicted in Fig. 3.2(a). Orient $\mathbb{S}^2$ so that "left" and "right" have the usual meaning.

Figure 3.4(a) is obtained after adding the neighbouring curves to the diagram. If the neighbouring points are labeled in such a way that related points have the same label, we obtain Fig. 3.4(b). Proceeding in the same way with the diagrams of Figs. 3.2(b), (d) and (f) we obtain Fig. 3.5.

Let A be a neighbouring point of $\mathcal{D}$, where the curves $\alpha \in \mathcal{D}$ and $\lambda \in \mathcal{N}(\mathcal{D})$ intersect. If τA is the neighbouring point of $\mathcal{D}$ related with A, then τA is the intersection of $\tau\alpha$ with a neighbouring curve $\mu \in \mathcal{N}(\mathcal{D})$. In this case we say that the neighbouring curves λ and μ have an *elementary relation of the first kind*. If α and $\tau\alpha$ are sister curves of $\mathcal{D}$ we say that the neighbouring curves α^- and $(\tau\alpha)^+$ have a *elementary relation of the second kind*. Equivalently, α^+ and $(\tau\alpha)^-$ also have an elementary relation of the second kind. Two neighbouring curves are *elementary related* if they have an elementary relation of the first or second kind.

Definition 3.13 (Classes of neighbouring curves). Two neighbouring curves λ and λ' of $\mathcal{D}$ are *in the same class* if there is a chain of neighbouring curves $\lambda = \lambda_0, \lambda_1, \ldots, \lambda_{m-1}, \lambda_m = \lambda'$ such that λ_i and λ_{i+1} are elementary related for all $i = 0, \ldots, m-1$.

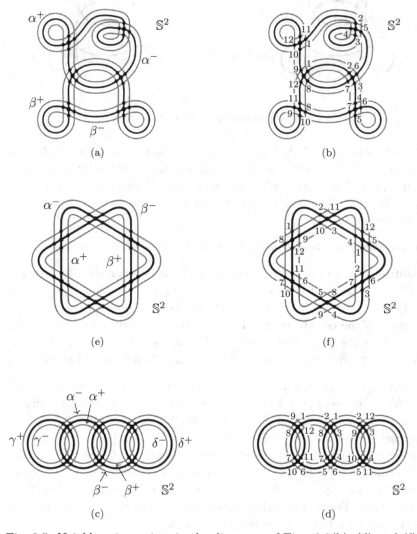

Fig. 3.5: Neighbouring points in the diagrams of Figs. 3.2(b), (d) and (f).

Notice the symmetry in the definition above:

Proposition 3.14. *If λ, μ are neighbouring curves of $\mathcal{D}$ and $\lambda', \mu' \in \mathcal{N}(\mathcal{D})$ are their respective opposite curves in $\mathcal{N}(\mathcal{D})$, then λ and μ are in the same class if and only if λ' and μ' are in the same class.*

Proof. It is enough to check that if λ and μ have an elementary relation of any kind λ' and μ' also have an elementary relation of the same kind, but this follows from the definition of elementary relations. □

Now we are able to state the Johansson Theorem:

Theorem 3.15 ([37]). *A diagram is realizable if and only if it has no opposite neighbouring curves in the same class.*

In [10], Carter proves a similar result for slightly different immersions to those studied in this work.

Example 3.16. In Fig. 3.4, the neighbouring curve α^- contains the neighbouring points 4, 3, 3, 12, 12 and 11, α^+ passes through the points 5, 2, 2, 1, 1 and 10, the curve β^- meets the points 6, 6, 5, 10, 9 and 9, and finally β^+ contains the points 7, 7, 4, 11, 8 and 8. Then, α^- and β^+ have elementary relations of the first and second kind, and the same happens with α^+ and β^-. Hence, there are two classes of curves, $\{\alpha^-, \beta^+\}$ and $\{\alpha^+, \beta^-\}$, and by Theorem 3.15 the diagram is realizable.

In the diagram of Fig. 3.5(a), the curve α^- passes through 5, 2, 3, 12, 12 and 11, and the curve α^+ meets 4, 3, 2, 1, 1 and 10. Therefore, α^- and α^+ have an elementary relation of the first kind and the diagram cannot be realized.

By applying Theorem 3.15 to the rest of the diagrams of Example 3.3, we see that all of them are realizable except that of Fig. 3.2(b).

Let M be a 3-manifold and let $f : S \to M$ be a parametrization of a Dehn surface Σ. Since both S and M are orientable, using a normal vector field, it is possible to construct a *thickening of f*, that is, an immersion

$$F : S \times [-1, 1] \to M$$

such that $F(P, 0) = f(P)$ for all $P \in S$. The thickening can be defined such that the parallel surfaces $\Sigma^- = F(S \times \{-1\})$ and $\Sigma^+ = F(S \times \{1\})$ are transverse and the configuration of Σ^-, Σ and Σ^+ around a simple, double or triple point of Σ can be chosen to be like those depicted in Fig. 3.6. Moreover, we can assume that $|\mathcal{N}(\mathcal{D})|$ agrees with $f^{-1}(\Sigma^- \cup \Sigma^+)$. In this

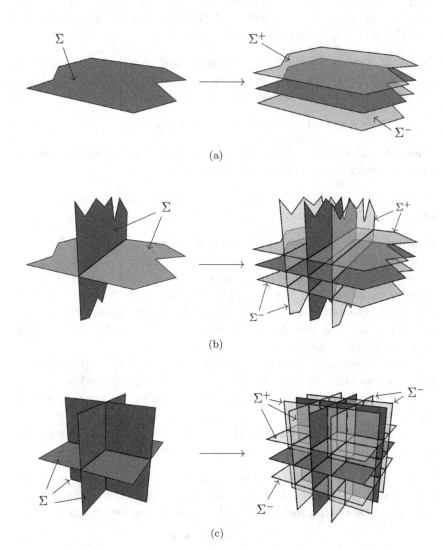

Fig. 3.6: Configuration of Σ, Σ^- and Σ^+ near (a) a simple point, (b) a double point, or (c) a triple point of Σ.

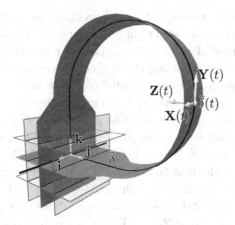

Fig. 3.7: Three-dimensional pieces.

way, the images under f of the neighbouring points of $\mathcal{D}$ are the intersection points of the double curves of Σ with $\Sigma^- \cup \Sigma^+$.

In this situation, a neighbouring curve of $\mathcal{D}$ is *in the bottom sheet* if its image under f is contained in Σ^-, and that it is *in the top sheet* if its image under f is contained in Σ^+. It is clear that opposite curves of $\mathcal{N}(\mathcal{D})$ are in opposite sheets. It is also straightforward to see that two neighbouring curves of $\mathcal{D}$ having an elementary relation of the first kind are in the same sheet.

Proof of Theorem 3.15. Assume that $\mathcal{D}$ is realizable. Let $f : S \to M$ be a transverse immersion realizing $\mathcal{D}$, where M is a 3-manifold. Take a thickening F of f as above.

Consider the Euclidean space $\mathbb{R}^3$ with the standard orientation given by the canonical basis $\{\mathbf{i}, \mathbf{j}, \mathbf{k}\}$. Imagine first that Σ is immersed in $\mathbb{R}^3$ in such a way that the following holds:

(Q1) The intersection of the cube $Q = [-2, 2]^3$ with Σ coincides with the intersection of Q with the union of the planes $y = 0$ and $z = 0$.

(Q2) The intersection of Q with Σ^- coincides with the intersection of Q with the union of the planes $y = -1$ and $z = -1$.

(Q3) The intersection of Q with Σ^+ coincides with the intersection of Q with the union of the planes $y = 1$ and $z = 1$.

(Q4) At the origin $\mathbf{O} = (0, 0, 0)$ the positive orientation of Σ on the sheet contained in $z = 0$ is that given by the frame $\{\mathbf{i}, \mathbf{j}\}$.

Let $O, \tau O \in S$ be the points of S in the preimage by f of the origin $\mathbf{O}$, so chosen that a neighbourhood of O in S is mapped by f into the plane $z = 0$ and a neighbourhood of τO in S is mapped by f into the plane $y = 0$. Let $\alpha, \tau\alpha$ be the curves of $\mathcal{D}$ passing through $O, \tau O$, respectively. If the double curve of Σ passing through $\mathbf{O}$ is oriented by the vector $\mathbf{i}$, by condition (Q4) $f \circ \alpha^-$ passes through the cube Q following the line $z = 0, y = 1$. That is, α^- is in the top sheet. Let $\delta : I \to S$ be a path from O to τO and consider its image closed path $\bar{\delta} = f \circ \delta$ in M. The reference frame $\{\mathbf{i}, \mathbf{j}, \mathbf{k}\}$ can be transported continuously along $\bar{\delta}$ using three maps $\mathbf{X}, \mathbf{Y}, \mathbf{Z} : I \to \mathbb{R}^3$ such that for each $t \in I$ the first two vectors $\{\mathbf{X}(t), \mathbf{Y}(t)\}$ form an orthogonal basis of $df(T_{\delta(t)}S)$ and $\mathbf{Z}(t)$ is orthogonal to $\mathbf{X}(t)$ and $\mathbf{Y}(t)$. This last condition implies that $\mathbf{Z}(t)$ "points toward" Σ^+ for all $t \in I$ as $\mathbf{Z}(0)$ does. Hence, we can assume that $\mathbf{Z}(1) = \mathbf{j}$. If it is required that $\mathbf{X}(1) = \mathbf{i}$, then $\mathbf{Y}(1) = \pm\mathbf{k}$. By continuity, $\{\mathbf{X}(t), \mathbf{Y}(t), \mathbf{Z}(t)\}$ is a positive basis of $\mathbb{R}^3$ for all $t \in I$, it turns out that $\mathbf{Y}(1) = -\mathbf{k}$. Definition 3.11 implies that α_2^- is in the bottom sheet. Using the orientability of S and M and some elementary riemannian geometry, this construction can be adapter to a generic Σ in a generic 3-manifold M. Hence, *two curves of $\mathcal{N}(\mathcal{D})$ having an elementary relation of the second kind are in the same sheet.* Thus, if two neighbouring curves belong to the same class, they must be in the same sheet. Since f is 2-sided, opposite neighbouring curves cannot be in the same sheet and therefore they cannot be in the same class.

Assume now that $\mathcal{N}(\mathcal{D})$ has no opposite curves in the same class. By Proposition 3.14 any class Λ of $\mathcal{D}$ has an *opposite class* Λ' composed by the curves of $\mathcal{N}(\mathcal{D})$ opposite to those in Λ. The set of classes of $\mathcal{N}(\mathcal{D})$ can be split into two sets, the *bottom group* and the *top group*, such that opposite classes belong to different groups. Such splitting of the classes of $\mathcal{N}(\mathcal{D})$ induces a splitting of $\mathcal{N}(\mathcal{D})$ which is called a *partition* of $\mathcal{N}(\mathcal{D})$. Note that any partition $\mathcal{P}$ of $\mathcal{N}(\mathcal{D})$ has an *opposite partition* $\mathcal{P}'$ which is obtained by swapping the bottom and top groups.

Fix a partition $\mathcal{P}$ of $\mathcal{N}(\mathcal{D})$. If we cut S along the curves of $\mathcal{N}(\mathcal{D})$, we obtain a collection of 2-*dimensional pieces*, each of which is the closure of a connected component of $S - |\mathcal{N}(\mathcal{D})|$. These pieces are of three types:

(A) type A pieces are *squares* with a crossing of $\mathcal{D}$ in their interior;
(B) type B pieces are *rectangles* or *annuli* containing exactly one arc or a complete curve of $\mathcal{D}$, respectively, without crossings; and
(C) type C pieces are free form ones. Each of them is contained in a connected component of $S - |\mathcal{D}|$.

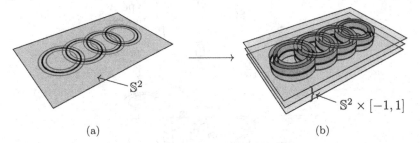

Fig. 3.8: Thickening the diagram.

Now, consider the product manifold $S \times [-1, 1]$ and the immersed surface $|\mathcal{N}(\mathcal{D})| \times [-1, 1]$ in it (see Fig. 3.8). Let $e : S \to S \times [-1, 1]$ be the embedding defined by $e(x) = (x, 0)$ for each $x \in S$. We sometimes identify objects in S with their images under e.

Cutting $S \times [-1, 1]$ along $|\mathcal{N}(\mathcal{D})| \times [-1, 1]$, we obtain a collection of 3-*dimensional pieces*. Each of them is the product of one 2-dimensional piece with $[-1, 1]$. A 3-dimensional piece is of type A, B or C if it is the product of a 2-dimensional piece of type A, B or C with $[-1, 1]$, respectively (see Fig. 3.9). The boundary of these pieces have a cornered structure that allows to talk about their *faces*, *edges* and *vertices* in a natural way.

For a 3-dimensional piece W, its face contained in $S \times \{-1\}$ is the *bottom horizontal face* of W, and its face contained in $S \times \{1\}$ is the *top horizontal face* of W. The rest of faces of W have the form $w \times [-1, 1]$, where w is an arc of neighbouring curve of $\mathcal{D}$ or a complete neighbouring curve. If w is contained in a neighbouring curve of $\mathcal{D}$ of the bottom (resp. top) group, the face $w \times [-1, 1]$ of W is a *bottom vertical face* (resp. *top vertical face*) of W. If W is of type A or B, there is at least one curve $\alpha \in \mathcal{D}$ passing through W. If $|\alpha|$ has no crossings of $\mathcal{D}$, it is completely contained in W. In this case, W is a *torus-like* type B piece, i.e., homeomorphic to a solid torus, and it only has four annular faces. If α passes through at least one crossing of $\mathcal{D}$, there is at least one arc $a \subset |\alpha|$ contained in W with its endpoints in ∂W. The arc a is located between two consecutive neighbouring points of $\mathcal{D}$ lying in $|\alpha|$, and it is an *axis* of W. Faces of W intersecting a are called *orthogonal* to a, and vertical faces of W not intersecting a are called *parallel* to a. The arc a has a *sister arc* τa, contained in $|\tau\alpha|$, whose points are related by $\mathcal{D}$ with those of a. The 3-dimensional piece that contains τa is the *sister piece* of W with respect to a. Each type B piece only has one axis and one sister piece, while a type A piece has two axes and two sister pieces.

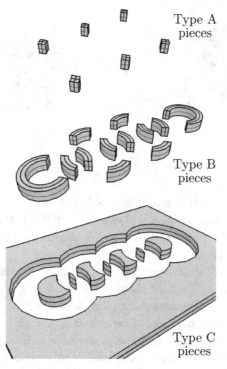

Fig. 3.9: Three-dimensional pieces.

Let $\mathcal{Y}$ be a type B 3-dimensional piece of $S \times [-1, 1]$ which is not torus-like, and let b be its axis. Let $\tau\mathcal{Y}$ be the sister piece of $\mathcal{Y}$, whose axis is τb. Take a diffeomorphism $p_{\mathcal{Y}} : \mathcal{Y} \to \tau\mathcal{Y}$ such that:

(i) it maps each point of b to its related point in τb;

(ii) it maps the bottom (resp. top) horizontal face of $\mathcal{Y}$ onto the bottom (resp. top) vertical face of $\tau\mathcal{Y}$ parallel to τb; and

(iii) it maps the bottom (resp. top) vertical face of $\mathcal{Y}$ parallel to b onto the bottom (resp. top) horizontal face of $\tau\mathcal{Y}$.

As a consequence, $p_{\mathcal{Y}}$ also maps the faces of $\mathcal{Y}$ orthogonal to b onto the faces of $\tau\mathcal{Y}$ orthogonal to τb. Such $p_{\mathcal{Y}}$ is the *identifying map* of $\mathcal{Y}$. The elementary relation of the second kind guarantees that $p_{\mathcal{Y}}$ is orientation-preserving.

If $\mathcal{Y}$ is a torus-like type B piece of $S \times [-1, 1]$ and β is the curve of $\mathcal{D}$ contained in $\mathcal{Y}$, the piece $\tau\mathcal{Y}$ that contains $\tau\beta$ also is torus-like. The identifying map of $\mathcal{Y}$ is a diffeomorphism $p_{\mathcal{Y}} : \mathcal{Y} \to \tau\mathcal{Y}$ such that:

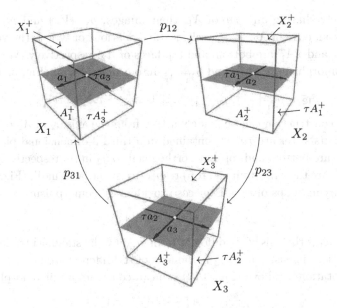

Fig. 3.10: Identification maps between type A pieces.

(i) it maps each point of $|\beta|$ to its related point in $|\tau\beta|$;

(ii) it maps the bottom (resp. top) horizontal face of Y onto the bottom (resp. top) vertical face of τY; and

(iii) it maps the bottom (resp. top) vertical face of $\mathcal{Y}$ onto the bottom (resp. top) horizontal face of $\tau\mathcal{Y}$.

We repeat these constructions for each type B piece of $S \times [-1, 1]$, imposing that identifying maps of sister pieces are inverse to each other, that is, $p_{\tau \mathcal{Y}} = p_{\mathcal{Y}}^{-1}$ for every type B piece $\mathcal{Y}$.

Let $\mathcal{X}_1$ be a 3-dimensional type A piece of $S \times [-1, 1]$ and let a_1 be one of its axes. The sister arc τa_1 of a_1 is an axis of another 3-dimensional type A piece $\mathcal{X}_2$ of $S \times [-1, 1]$. If $\mathcal{Y}_1^-$ and $\mathcal{Y}_1^+$ are the type B pieces neighbours of $\mathcal{X}_1$ at the endpoints of a_1, they share the faces A_1^- and A_1^+ with $\mathcal{X}_1$, respectively, orthogonal to a_1. We assume that A_1^- is a bottom face and that A_1^+ is a top face (Fig. 3.10). The identifying map $p_{\mathcal{Y}_1^-}$ maps A_1^- diffeomorphically onto a face τA_1^- of $\mathcal{X}_2$, its *sister face*, which is a vertical face of $\mathcal{X}_2$ orthogonal to τa_1. Equivalently, $p_{\mathcal{Y}_1^+}$ maps A_1^+ onto τA_1^+, where τA_1^+ denotes the opposite face of τA_1^- in $\mathcal{X}_2$. The conditions imposed to the identifying maps of type B pieces imply that if V_- and V_+ are vertices of A_1^- and A_1^+

respectively sharing an edge of $\mathcal{X}_1$, their images, $p_{y_1^-}(V_-)$ and $p_{y_1^+}(V_+)$, also share an edge of $\mathcal{X}_2$. By the elementary relations of type I, the vertical faces τA_1^- and τA_1^+ are bottom and top faces of $\mathcal{X}_2$, respectively. We group the diffeomorphism $p_{y_1^-}|_{A_1^-}$ and $p_{y_1^+}|_{A_1^+}$ into a unique map (Fig. 3.10)

$$\tilde{p}_{12} = p_{y_1^-}|_{A_1^-} \cup p_{y_1^+}|_{A_1^+} : A_1^- \cup A_1^+ \to \tau A_1^- \cup \tau A_1^+.$$

A similar construction can be made starting from the axis a_2 of $\mathcal{X}_2$ different from τa_1: its sister arc τa_2 is contained in a third 3-dimensional piece $\mathcal{X}_3$; if A_2^-, A_2^+ are bottom and top faces orthogonal to a_2 in $\mathcal{X}_2$ respectively, and $\tau A_2^-, \tau A_2^+$ are their (bottom and top) respective sister faces in $\mathcal{X}_3$ (Fig. 3.10), the identifying maps of type B pieces provide a diffeomorphism

$$\tilde{p}_{23} : A_2^- \cup A_2^+ \to \tau A_2^- \cup \tau A_2^+.$$

Finally, if a_3 is the axis of $\mathcal{X}_3$ different from τa_2, by the standard triple point configuration, its sister arc τa_3 is the axis of $\mathcal{X}_1$ different from a_1. Using a similar notation as above (Fig. 3.10), we can construct a diffeomorphism

$$\tilde{p}_{31} : A_3^- \cup A_3^+ \to \tau A_3^- \cup \tau A_3^+,$$

mapping the bottom and top faces A_3^- and A_3^+ of $\mathcal{X}_3$ orthogonal to a_3 onto their sister faces τA_3^- and τA_3^+, respectively, in $\mathcal{X}_1$.

Claim 3.17. *If $\mathcal{V}_1$ is the set of vertices of $\mathcal{X}_1$, then*

$$(\tilde{p}_{31} \circ \tilde{p}_{23} \circ \tilde{p}_{12})\,|_{\mathcal{V}_1} = \mathrm{id}_{\mathcal{V}_1}.$$

Proof of Claim 3.17. Consider an index $i = 1, 2, 3$. The indices $i - 1$ and $i + 1$ are considered modulo 3. We identify each vertex V of $\mathcal{X}_i$ with a triple of *coordinates* $(\varepsilon_1, \varepsilon_2, \varepsilon_3)$, with $\varepsilon_j = \pm 1$ for $j = 1, 2, 3$. The first coordinate ε_1 is given by the axis a_i: $\varepsilon_1 = -1$ if V is in A_i^-, and $\varepsilon_1 = 1$ if $V \in A_i^+$. In the same way, $\varepsilon_2 = -1$ if $V \in \tau A_{i-1}^-$, and $\varepsilon_2 = 1$ if $V \in \tau A_{i-1}^+$. Finally, we take $\varepsilon_3 = -1$ if V is in the bottom horizontal face of $\mathcal{X}_i$ and $\varepsilon_3 = 1$ if V is in the top horizontal face of $\mathcal{X}_i$. By the properties of the gluing maps of type B pieces, if the vertex $V \in \mathcal{X}_i$ has coordinates $(\varepsilon_1, \varepsilon_2, \varepsilon_3)$ in $\mathcal{X}_i$, $\tilde{p}_{i(i+1)}(V)$ has coordinates $(\varepsilon_3, \varepsilon_1, \varepsilon_2)$ in $\mathcal{X}_{i+1}$. The proof is now straightforward. $\qquad\square$

If a is an axis of $\mathcal{X}_i$, for some $i = 1, 2, 3$, the symbol a'' denotes the union of the four edges of $\mathcal{X}_i$ parallel to a. The map $\tilde{p}_{31}^{-1}$ sends a_1'' onto the four vertical edges of $\mathcal{X}_3$, and these are mapped by $\tilde{p}_{23}^{-1}$ onto $(\tau a_1)''$. By the previous claim, since $\tilde{p}_{12}$ and $\tilde{p}_{23}^{-1} \circ \tilde{p}_{31}^{-1}$ agree over $\mathcal{V}$, there is a diffeomorphism $p_{12} : \mathcal{X}_1 \to \mathcal{X}_2$ that agrees with $\tilde{p}_{12}$ over $A_1^- \cup A_1^+$ and with $\tilde{p}_{23}^{-1} \circ \tilde{p}_{31}^{-1}$ over a_1''.

At this step, we have two maps sending subsets of $\mathcal{X}_2$ to subsets of $\mathcal{X}_3$: $\tilde{p}_{23}$ and $\tilde{p}_{31}^{-1} \circ p_{12}^{-1}$. If we call X_2^- and X_2^+ the bottom and top horizontal faces of $\mathcal{X}_2$, respectively, the maximal subset of $\mathcal{X}_2$ over which $\tilde{p}_{31}^{-1} \circ p_{12}^{-1}$ can be defined is $X_2^- \cup X_2^+$, and the intersection of $X_2^- \cup X_2^+$ with $A_2^- \cup A_2^+$ is $\tau a_1''$. Since, by the construction of p_{12}, the maps $\tilde{p}_{23}$ and $\tilde{p}_{31}^{-1} \circ p_{12}^{-1}$ agree over $\tau a_1''$, we can construct a diffeomorphism $p_{23} : \mathcal{X}_2 \to \mathcal{X}_3$ that agrees with $\tilde{p}_{23}$ over $A_2^- \cup A_2^+$ and with $\tilde{p}_{31}^{-1} \circ p_{12}^{-1}$ over $X_2^- \cup X_2^+$. Defining $p_{31} = p_{12}^{-1} \circ p_{23}^{-1}$, by construction the map p_{31} agrees with $\tilde{p}_{31}$ over $A_3^- \cup A_3^+$. Finally, we define the maps p_{21}, p_{32}, p_{13} as the inverse maps of p_{12}, p_{23}, p_{31}, respectively. The *identifying map p_{ij} from $\mathcal{X}_i$ to $\mathcal{X}_j$* is orientation-preserving, for $i, j = 1, 2, 3$ and $i \neq j$.

We repeat this construction for all the type A pieces of $S \times [-1, 1]$. Let $\mathcal{I}$ be the set of all identifying maps (between type B or type A pieces) so constructed. Given $x, y \in S \times [-1, 1]$

$$x \sim_{\mathcal{I}} y \iff x = y \text{ or there exists } p \in \mathcal{I} \text{ with } p(x) = y.$$

This relation is symmetric and, by the construction of the identifying maps between type A pieces, it is also reflexive. The relation $\sim_{\mathcal{I}}$ is indeed an equivalence relation on $S \times [-1, 1]$. The equivalence class of a point $x \in S \times [-1, 1]$ has 3 points if x lies in a type A piece, 2 points if x is in a type B piece but not in a type A piece, or 1 point if x lies in a type C piece but not in a type B piece. The equivalence relation $\sim_{\mathcal{I}}$ agrees with the equivalence relation $\sim_{\mathcal{D}}$ when restricted to S, and the quotient space

$$\widehat{M}(\mathcal{D}) = S \times [-1, 1]/\sim_{\mathcal{I}}$$

is an orientable 3-manifold with boundary and corners that, gluing 3-disks or handlebodies to its boundary components, it can be thought as a subset of a closed 3-manifold M. If $\pi : S \times [-1, 1] \to \widehat{M}(\mathcal{D})$ is the natural projection and $\iota : \widehat{M}(\mathcal{D}) \hookrightarrow M$ is the inclusion map, then $\iota \circ \pi \circ e : S \to M$ is an immersion realizing $\mathcal{D}$. $\qquad\square$

3.4 Filling diagrams

Let $\mathcal{D}$ be a realizable diagram in S and let $f : S \to M$ be an immersion realizing $\mathcal{D}$, where S is a closed orientable surface. The following result can be deduced from the proof of the Johansson Theorem.

Proposition 3.18. *The number of classes of neighbouring curves of $\mathcal{D}$ is twice the number of connected components of* $\mathrm{Sing}(f)$. *In particular, if $\mathcal{D}$ is connected, it has exactly two classes of neighbouring curves.*

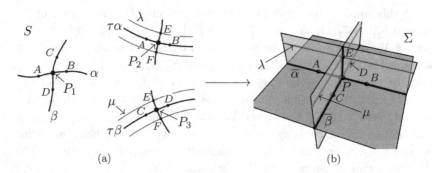

(a)   (b)

Fig. 3.11: Type I relation viewed in Σ.

Proof. Consider two curves α and $\beta \in \mathcal{D}$. A neighbouring curve of α has an elementary relation of type II with a neighbouring curve of β if and only if $\beta = \tau\alpha$. A neighbouring curve of α has an elementary relation of type I with a neighbouring curve of β if and only if $|f \circ \alpha|$ and $|f \circ \beta|$ intersect transversely at a triple point of f (Fig. 3.11). It follows that α and β have neighbouring curves in the same class if and only if $|f \circ \alpha|$ and $|f \circ \beta|$ are in the same connected component of $\mathrm{Sing}(f)$. Since $\mathcal{D}$ is realizable, opposite neighbouring curves lie in different classes and there are exactly two classes of neighbouring curves for each connected component of $\mathrm{Sing}(f)$. If $\mathcal{D}$ is connected, $\mathrm{Sing}(f)$ is also connected. $\square$

If $\mathcal{D}$ is connected, the steps in the proof of the Johansson Theorem from the diagram $\mathcal{D}$ until the construction of the 3-manifold with boundary $\widehat{M}(\mathcal{D})$ are "unique". Since there are only two classes of neighbouring curves, the partition $\mathcal{P}$ is unique up to renaming of the bottom and top groups. The set of neighbouring curves $\mathcal{N}(\mathcal{D})$ is unique up to ambient isotopy of S. Hence, the 3-dimensional pieces of $S \times [-1,1]$ are unique up to ambient isotopy of $S \times [-1,1]$. The identifying maps between type B pieces are unique up to isotopy, so they are the identifying maps between type A pieces. Therefore:

Theorem 3.19. *If the diagram $\mathcal{D}$ is connected, the manifold $\widehat{M}(\mathcal{D})$ built from $\mathcal{D}$ as in the proof of the Johansson Theorem is unique up to diffeomorphism.* $\square$

Moreover, using tubular neighbourhoods, it is easy to prove the following result:

Lemma 3.20. *If the diagram $\mathcal{D}$ is connected, there is a regular neighbourhood of $f(S)$ in M diffeomorphic to $\widehat{M}(\mathcal{D})$.* $\square$

Theorem 3.21. *The realizable diagram $\mathcal{D}$ is filling if and only if it fills S and $\partial \widehat{M}(\mathcal{D})$ is a disjoint union of 2-spheres.*

Proof. Assume first that $\mathcal{D}$ and f are filling. By Proposition 3.8, $\mathcal{D}$ fills S. By Lemma 3.20, there is a regular neighbourhood $\widetilde{N}$ of $f(S)$ in M homeomorphic to $\widehat{M}(\mathcal{D})$. Since f is filling, $\partial \widetilde{N}$ is a disjoint union of 2-spheres, and therefore, $\partial \widehat{M}(\mathcal{D})$ is a disjoint union of 2-spheres.

Finally, assume that $\mathcal{D}$ fills S and that $\partial \widehat{M}(\mathcal{D})$ is a disjoint union of 2-spheres. Let M be the closed 3-manifold obtained by gluing a 3-disk along each connected component of $\partial \widehat{M}(\mathcal{D})$. The map $\iota \circ \pi \circ e : S \to M$, as in the proof of the Johansson Theorem, is a filling immersion realizing $\mathcal{D}$. $\quad\square$

Definition 3.22. If $\mathcal{D}$ is filling, define $M(\mathcal{D})$ as the 3-manifold obtained by gluing a 3-disk along each connected component of $\partial \widehat{M}(\mathcal{D})$.

When the diagram $\mathcal{D}$ is filling, the manifold $M(\mathcal{D})$ also is unique, in the sense of the following result:

Theorem 3.23. *If $\mathcal{D}$ and f are filling, then M is homeomorphic to $M(\mathcal{D})$.*

Proof. In the previous proof, we have seen that a regular neighbourhood $\widetilde{N}$ of $f(S)$ is homeomorphic to $\widehat{M}(\mathcal{D})$ and, since f is filling, the closure of each connected component of $M - \widetilde{N}$ is a 3-disk. Therefore, M is obtained by gluing disks along their boundary to the boundary of $\widetilde{N}$, which implies $M(\mathcal{D})$ and M are homeomorphic. $\quad\square$

The result above, together with Theorem 2.11, means that Johansson diagrams of filling Dehn spheres allows to describe any 3-manifold.

Definition 3.24. The Johansson diagram of a filling Dehn surface of a 3-manifold M is a *Johansson representation* of M or a *Johansson representation in S* of M.

It is interesting to obtain practical methods for checking when a realizable diagram is filling.

Assume that $\mathcal{D}$ is connected and fills S. Take a partition $\mathcal{P}$ of the set of adjacent curves of $\mathcal{D}$, and recall the construction of $\widehat{M}(\mathcal{D})$ as a quotient of $S \times [-1, 1]$ in the proof of the Johansson Theorem. Take the natural projection $\pi : S \times [-1, 1] \to \widehat{M}(\mathcal{D})$ as there. Put $f = \pi \circ e$ and $\Sigma = f(S)$. For simplicity, let us denote the 3-manifold with boundary $\widehat{M}(\mathcal{D})$ just by $\widehat{M}$.

Let $C_1, \ldots, C_m$ be the type C 2-dimensional pieces of S, and for every $i = 1, \ldots, m$ define the *parallel faces* to C_i as the sets $C_i^- = C_i \times \{-1\}$ and $C_i^+ = C_i \times \{1\}$. Let b be an arc of $|\mathcal{D}|$ which is the axis of a 3-dimensional type B piece $\mathcal{Y}$ of $S \times [-1, 1]$, and let τb be its sister arc, which is the axis of the type B piece $\tau \mathcal{Y}$. Let b_-, b_+ be the two arcs of neighbouring curves that run parallel to b at both sides of b in $\mathcal{Y}$, so chosen that b_- belongs to an adjacent curve of the bottom group and b_+ belongs to an adjacent curve of the top group. With the same criterion, denote the two arcs of neighbouring curves that run parallel to τb at both sides of τb in $\tau \mathcal{Y}$ by $(\tau b)_-$ and $(\tau b)_+$. To avoid confusion, it should be noted that the subscripts $\pm$ used here for the neighbouring arcs to b and τb refer to *top-bottom* instead of *right-left*, unlike the superscripts $\pm$ used before for denoting the neighbouring curves of the curves of $\mathcal{D}$. Let C_1, C_2, C_3, C_4 be the 2-dimensional type C pieces which are incident with $b_-, b_+, \tau b_-, \tau b_+$, respectively. By the construction of the identifying map $p_{\mathcal{Y}}$, the parallel faces of $C_1, \ldots, C_4$ become glued around $f(b)$ along their edges parallel to b or τb, respectively, according to the following scheme (see Fig. 3.12):

$$C_1^- \text{ is adjacent to } C_3^-,$$
$$C_1^+ \text{ is adjacent to } C_4^-,$$
$$C_2^- \text{ is adjacent to } C_3^+,$$
$$C_2^+ \text{ is adjacent to } C_4^+.$$

This allows to construct all the connected components of $\partial \widehat{M}$ out from $\mathcal{D}$.

The number k of type A pieces agrees with the number of crossings of $\mathcal{D}$, the number ℓ of type B pieces is the same as the number of connected components of $|\mathcal{D}| - \{\text{crossings of } \mathcal{D}\}$, and the number m of type C pieces coincides with the number of connected components of $S - |\mathcal{D}|$. The diagram $\mathcal{D}$ fills S, therefore

$$k - \ell + m = \chi(S).$$

Since $|\mathcal{D}|$ is a 4-valent graph $\ell = 2k$, and hence $m = k + \chi(S)$. The image under π of the parallel faces of $C_1, \ldots, C_m$ are the 2-cells of a cellular decomposition of $\partial \widehat{M}$. This cellular decomposition has eight vertices around each triple point of Σ, four 1-cells (edges) around each edge of Σ, and two 2-cells (faces) around each face of Σ.

Since $\mathcal{D}$ has k crossings, Σ has $\frac{k}{3}$ triple points and $\partial \widehat{M}$ has $8\frac{k}{3}$ vertices. The number of edges of Σ is $\frac{\ell}{2} = k$, therefore $\partial \widehat{M}$ has $4k$ edges. The Euler-Poincaré characteristic of $\partial \widehat{M}$ is

$$\chi(\partial \widehat{M}) = 8\frac{k}{3} - 4k + 2(k + \chi(S)) = 2\left(\frac{k}{3} + \chi(S)\right) = 2(\# \operatorname{T}(\Sigma) + \chi(S)). \quad (3.1)$$

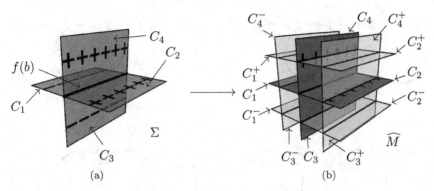

Fig. 3.12: The construction of $\widehat{M}$.

If the boundary of $\widehat{M}$ is composed by n 2-spheres, then

$$n = \#\,\mathrm{T}(\Sigma) + \chi(S).$$

Theorem 3.25. *The diagram $\mathcal{D}$ is filling if and only if the set $\partial\widehat{M}$ has $\#\,\mathrm{T}(\Sigma) + \chi(S)$ connected components.*

Proof. As we have seen, if $\mathcal{D}$ is filling, then $\partial\widehat{M}$ has $\#\,\mathrm{T}(\Sigma) + \chi(S)$ connected components. On the other hand, if $\partial\widehat{M}$ has $n = \#\,\mathrm{T}(\Sigma) + \chi(S)$ connected components $S_1, \ldots, S_n$ and one of them is not a 2-sphere, the Euler characteristic of $\partial\widehat{M}$,

$$\chi(\partial\widehat{M}) = \chi(S_1) + \cdots + \chi(S_n),$$

is strictly less than $2(\#\,\mathrm{T}(\Sigma) + \chi(S))$, in contradiction with Eq. (3.1). $\square$

Thus, counting the number of connected components of $\widehat{M}$, without constructing them explicitly, it is possible to determine if a diagram is filling or not.

Example 3.26. Let us search the filling diagrams supported by the four-circles graph of Fig. 3.13. Assume that $\mathcal{D}$ is such a diagram.

It is clear that the two extremal curves of the diagram, each of which having two crossings, must be sister curves, and that the two middle curves of the diagram must be sisters too. The two middle curves of the diagram, say α and $\tau\alpha$, can be identified in the eight possible ways depicted in Fig. 3.13, where related neighbouring points are equally denoted. Among those eight possible ways, the only ones whose crossings can be grouped in triplets of pairwise related crossings are the diagrams of Figs. 3.13(a) and 3.13(g).

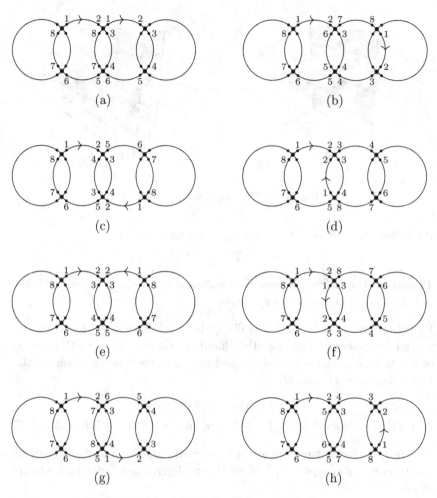

Fig. 3.13: Identifications of the middle curves.

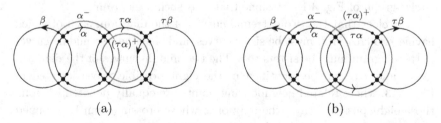

Fig. 3.14: Identifying the extremal curves.

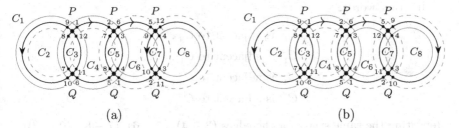

Fig. 3.15: The two classes of neighbouring curves and the type C pieces.

In these two diagrams, draw the neighbouring curves α^- and $\tau\alpha^+$, and choose two related crossings as the starting points of the extremal curves β and $\tau\beta$ of the diagram. In order to obtain a realizable diagram, it turns out that the curve β pointing towards α^- must be identified with the curve $\tau\beta$ pointing towards $(\tau\alpha)^+$ at their starting points. Therefore, in the diagrams of Figs. 3.13(a) and 3.13(g) the curves β and $\tau\beta$ must be identified in the way depicted in Fig. 3.14.

Adding the rest of neighbouring curves to the diagrams of Fig. 3.14, Fig. 3.15 is obtained, where all the neighbouring points of the diagram and the type C pieces $C_1 \ldots, C_8$ are labelled. Assume that the bottom group of neighbouring curves is composed by the (gray) curves α^-, $(\tau\alpha)^+$, β^+ and $(\tau\beta)^-$, while the top group is composed by their opposite neighbouring (dashed grey) curves: α^+, $(\tau\alpha)^-$, β^- and $(\tau\beta)^+$.

In the diagram of Fig. 3.15(a), if we look at the edge $(1-2)$ of α containing the neighbouring points 1 and 2, C_1 lies at the *bottom side* of $(1-2)$, while C_4 lies at the *top side* of $(1-2)$. On the other hand, if we look at the sister edge $(1-2)$ ying in $\tau\alpha$, C_6 lies at the bottom side of $(1-2)$ while C_1 lies at the top side of $(1-2)$. According to Fig. 3.12, the parallel faces of C_1, C_4 and C_6 become pasted in $\widehat{M}$ around $(1-2)$ by means of the following adjacency matrix:

			bottom		top	
	$(1-2)$	α	C_1		C_4	
		$\tau\alpha$	$-$	$+$	$-$	$+$
bottom	C_6	$-$	1	0	0	0
		$+$	0	0	1	0
top	C_1	$-$	0	1	0	0
		$+$	0	0	0	1

In other words,

$$C_1^- \text{ is adjacent to } C_6^-,$$
$$C_1^+ \text{ is adjacent to } C_1^-,$$
$$C_4^- \text{ is adjacent to } C_6^+,$$
$$C_4^+ \text{ is adjacent to } C_1^+.$$

Repeating the same study for the edges $(3-4)$, $(5-6)$, $(7-8)$, $(9-10)$ and $(11-12)$ we obtain the following adjacency matrices:

$(3-4)$	α	C_6		C_5	
$\tau\alpha$		$-$	$+$	$-$	$+$
C_7	$-$	1	0	0	0
	$+$	0	0	1	0
C_8	$-$	0	1	0	0
	$+$	0	0	0	1

$(5-6)$	α	C_1		C_4	
$\tau\alpha$		$-$	$+$	$-$	$+$
C_6	$-$	1	0	0	0
	$+$	0	0	1	0
C_1	$-$	0	1	0	0
	$+$	0	0	0	1

$(7-8)$	α	C_2		C_3	
$\tau\alpha$		$-$	$+$	$-$	$+$
C_5	$-$	1	0	0	0
	$+$	0	0	1	0
C_4	$-$	0	1	0	0
	$+$	0	0	0	1

$(9-10)$	β	C_1		C_2	
$\tau\beta$		$-$	$+$	$-$	$+$
C_7	$-$	1	0	0	0
	$+$	0	0	1	0
C_6	$-$	0	1	0	0
	$+$	0	0	0	1

$(11-12)$	β	C_4		C_3	
$\tau\beta$		$-$	$+$	$-$	$+$
C_8	$-$	1	0	0	0
	$+$	0	0	1	0
C_1	$-$	0	1	0	0
	$+$	0	0	0	1

With these adjacent matrices it turns out that $\partial\widehat{M}$ has four connected components, each of them containing one of the following sets of parallel faces to $C_1, \ldots, C_8$:

$$\{C_1^-, C_1^+, C_3^+, C_4^+, C_6^-, C_7^-\},$$
$$\{C_2^-, C_5^-, C_7^+\},$$
$$\{C_2^+, C_4^-, C_6^+, C_8^-\},$$
$$\{C_3^-, C_5^+, C_8^+\}.$$

By Theorem 3.25, the diagram of Fig. 3.15(a) is filling.

The same study for the diagram of Fig. 3.15(b) gives:

$(1-2)$	α	C_1		C_4	
$\tau\alpha$		−	+	−	+
C_1	−	1	0	0	0
	+	0	0	1	0
C_6	−	0	1	0	0
	+	0	0	0	1

$(3-4)$	α	C_6		C_5	
$\tau\alpha$		−	+	−	+
C_8	−	1	0	0	0
	+	0	0	1	0
C_7	−	0	1	0	0
	+	0	0	0	1

$(5-6)$	α	C_1		C_4	
$\tau\alpha$		−	+	−	+
C_1	−	1	0	0	0
	+	0	0	1	0
C_6	−	0	1	0	0
	+	0	0	0	1

$(7-8)$	α	C_2		C_3	
$\tau\alpha$		−	+	−	+
C_4	−	1	0	0	0
	+	0	0	1	0
C_5	−	0	1	0	0
	+	0	0	0	1

$(9-10)$	β	C_1		C_2	
$\tau\alpha$		−	+	−	+
C_1	−	1	0	0	0
	+	0	0	1	0
C_8	−	0	1	0	0
	+	0	0	0	1

$(11-12)$	α	C_4		C_3	
$\tau\alpha$		−	+	−	+
C_6	−	1	0	0	0
	+	0	0	1	0
C_7	−	0	1	0	0
	+	0	0	0	1

The boundary of $\widehat{M}$ has five connected components, each of them containing one of the following sets of parallel faces to $C_1, \ldots, C_8$:

$$\{C_1^-\},$$
$$\{C_1^+, C_2^-, C_4^-, C_6^-, C_8^-\},$$
$$\{C_2^+, C_5^-, C_8^+\},$$
$$\{C_3^-, C_4^+, C_6^+, C_7^-\},$$
$$\{C_3^+, C_5^+, C_7^+\},$$

and the diagram of Fig. 3.15(b) is not filling.

Chapter 4

Fundamental group of a Dehn sphere

While the title refers to "Dehn spheres", the constructions in this chapter using a diagram $\mathcal{D}$ in $\mathbb{S}^2$ are independent of whether $\mathcal{D}$ is realizable or not. Hence, results obtained from them are valid for pseudo Dehn spheres (Remark 3.5).

All coverings maps considered in this section are *unbranched*. For further details on them see [45, 61].

4.1 Coverings of Dehn spheres

Let $\mathcal{D}$ be a diagram on $\mathbb{S}^2$, let Σ be the quotient space of $\mathbb{S}^2$ under the relation induced by the diagram $\mathcal{D}$ in $\mathbb{S}^2$, and let $f : \mathbb{S}^2 \to \Sigma$ be the canonical projection.

Let $p : \Sigma^* \to \Sigma$ be an n-sheeted covering map of Σ, with $n \in \mathbb{N} \cup \{\infty\}$. Let $\mathbf{n}$ be an index set with $\#\mathbf{n} = n$.

Recall that, if $g : N \to \Sigma$ is a continuous map from a topological space N, a *lift of g under p* is a continuous map $g^* : N \to \Sigma^*$ such that $p \circ g^* = g$.

Let X_0 be a simple point of Σ. The preimage $f^{-1}(X_0)$ is a single point X of $\mathbb{S}^2$, and $p^{-1}(X_0)$ is a set with n distinct points $\{X_i\}_{i \in \mathbf{n}} \subset \Sigma^*$. Since $\mathbb{S}^2$ is simply connected, there exists a unique lift $f_i : \mathbb{S}^2 \to \Sigma^*$ of f under p such that $f_i(X) = X_i$, for each $i \in \mathbf{n}$ [45]. In fact, those are the only possible lifts of f under p. For any $i \in \mathbf{n}$ set $\Sigma_i = f_i(\mathbb{S}^2)$, then

$$\Sigma^* = \bigcup_{i \in \mathbf{n}} \Sigma_i.$$

The simply connectedness of $\mathbb{S}^2$ also implies that the preimage under p of a double curve of Σ is a collection of n double curves in Σ^*. Each of these n double curves is the intersection of two Dehn spheres of the family $\{\Sigma_i\}_{i \in \mathbf{n}}$. If $\alpha : \mathbb{S}^1 \to \mathbb{S}^2$ is a curve of $\mathcal{D}$, for each $i \in \mathbf{n}$, the map $f_i \circ \alpha$ parametrizes a

51

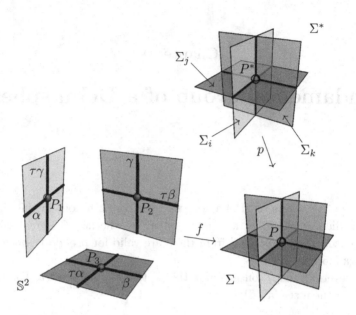

Fig. 4.1: The lift of a parametrization of a Dehn sphere.

double curve of Σ^*. There is $j \in \mathbf{n}$ such that Σ_i intersects Σ_j transversely at every point of $|f_i \circ \alpha|$. In this case Σ_j is said to be *transverse to Σ_i along* α. This defines a map

$$\theta\alpha : \mathbf{n} \to \mathbf{n}$$

by

$$\theta\alpha(i) = j,$$

if Σ_j is transverse to Σ_i along α. Obviously, if Σ_j is transverse to Σ_i along α, then Σ_i is transverse to Σ_j along $\tau\alpha$. Hence $\theta\tau\alpha = (\theta\alpha)^{-1}$. In particular, $\theta\alpha$ is a bijection. Therefore, it is an element of the group Ω_n of permutations of $\mathbf{n}$.

Let P be a triple point of Σ, and let P^* denote one of its n preimages under p in Σ^*. Assume that the 2-spheres Σ_i, Σ_j and Σ_k meet at P^* as in Fig. 4.1. Consider the three crossings P_1, P_2 and P_3 of $\mathcal{D}$ in the triplet of P, and label the curves of $\mathcal{D}$ meeting at them according to the standard triple point configuration (Fig. 4.1). In this situation,

$$j = \theta\alpha(i), \qquad\qquad k = \theta\beta(j), \qquad\qquad i = \theta\gamma(k),$$

therefore

$$(\theta\gamma \circ \theta\beta \circ \theta\alpha)(i) = i.$$

If P^* varies over all the preimages of P under p, we obtain that

$$\theta\gamma \circ \theta\beta \circ \theta\alpha = 1_n,$$

where $1_n \in \Omega_n$ is the identity permutation.

4.2 The diagram group

Consider the free group $\mathbb{F}_{\mathcal{D}}$ generated by the elements of $\mathcal{D}$. Let us consider the following relations in $\mathbb{F}_{\mathcal{D}}$:

Sister curves relations:

$$\alpha = \tau\alpha^{-1} \text{ for each } \alpha \in \mathcal{D};$$

Triple point relations: for any triplet $\{P_1, P_2, P_3\}$ of $\mathcal{D}$, if the curves of $\mathcal{D}$ intersecting at P_1, P_2, P_3 are labeled according to the standard triple point configuration (Fig. 3.1), then

$$\alpha\beta\gamma = 1.$$

Definition 4.1. Adding to $\mathbb{F}_{\mathcal{D}}$ the sister curves and triple point relations we obtain the *diagram group* $\pi(\mathcal{D})$ *of* $\mathcal{D}$.

Example 4.2. The two diagrams of Fig. 3.14 verify $\pi(\mathcal{D}) \approx \mathbb{Z}$.

Let Ω_n denote the group of permutations of $\mathbf{n}$ for $n \in \mathbb{N} \cup \{\infty\}$ as in previous section. Recall that a *representation* of a group G into Ω_n is a homomorphism $\rho : G \to \Omega_n$. A representation ρ is *faithful* if it is one-to-one, and it is *transitive* if for any pair $i, j \in \mathbf{n}$ there exists an element $a \in G$ such that $\rho(a)(i) = j$.

The relations introduced in $\pi(\mathcal{D})$ imply that each n-sheeted covering map p of Σ has an *associated representation* $\theta : \pi(\mathcal{D}) \to \Omega_n$ given by $\theta(\alpha) = \theta\alpha$. Since θ was constructed from the names $\{X_i\}_{i \in \mathbf{n}}$ given to the points of $p^{-1}(X_0)$, it is clear that θ is well defined up to "renaming", that is, up to conjugation with an element of Ω_n.

This relation between covering maps and representations in permutation groups is a property that $\pi(\mathcal{D})$ shares with the fundamental group $\pi_1(\Sigma)$ of Σ. It is well-known that the connected covering spaces of Σ are determined by transitive representations of $\pi_1(\Sigma)$ into the groups Ω_n (the *monodromy homomorphisms*) [61, Chap. VIII]. So it is natural to ask how $\pi(\mathcal{D})$ and $\pi_1(\Sigma)$ are related.

Theorem 4.3. *The groups $\pi(\mathcal{D})$ and $\pi_1(\Sigma)$ are isomorphic.*

The proof of this theorem will be delayed until Sec. 4.3. This result provides a presentation of the fundamental group of a Dehn or pseudo Dehn sphere Σ in terms of its Johansson diagram. Using the sister curves relations we can remove half of the generators, and in this way a ternary presentation of $\pi_1(\Sigma)$ is obtained (i.e., all relators are of length three). Thus, by Theorem 2.11, the fundamental group of any closed 3-manifold has a ternary presentation. A remark in [39, Problem 3.98] suggests that Theorem 4.3 was already known, probably by Haken, but we have not seen it in the literature until [69].

4.3   Coverings and representations

We have seen that each n-sheeted covering space of Σ has an associated representation $\theta : \pi(\mathcal{D}) \to \Omega_n$. The converse is also true:

Theorem 4.4. *Let $\theta : \pi(\mathcal{D}) \to \Omega_n$ be a representation. Then there exists an n-sheeted covering map $p : \widetilde{\Sigma} \to \Sigma$ such that θ is the associated representation of p. Moreover, the space $\widetilde{\Sigma}$ is connected if and only if θ is transitive.*

Proof. Given a curve $\alpha \in \mathcal{D}$, its n *copies* in $\mathbb{S}^2 \times \mathbf{n}$ are the maps

$$\alpha_i : \mathbb{S}^1 \to \mathbb{S}^2 \times \mathbf{n}$$

given by $\alpha_i(t) = (\alpha(t), i)$ with $t \in \mathbb{S}^1$, for $i \in \mathbf{n}$.

Consider the collection $\widetilde{\mathcal{D}}$ of the copies in $\mathbb{S}^2 \times \mathbf{n}$ of all the curves of $\mathcal{D}$. The representation θ allows to define a sistering $\widetilde{\tau}$ in $\widetilde{\mathcal{D}}$ as follows: given $\alpha \in \mathcal{D}$ and $i \in \mathbf{n}$, the sister curve of α_i in $\widetilde{\mathcal{D}}$ is the curve $(\tau\alpha)_{\theta\alpha(i)}$, where $\tau\alpha$ is the sister curve of α in $\mathcal{D}$.

Since θ is a homomorphism, the sister curves relations in $\pi(\mathcal{D})$ imply that $\theta\alpha = (\theta\tau\alpha)^{-1}$ for any $\alpha \in \mathcal{D}$. Therefore,

$$\widetilde{\tau}(\widetilde{\tau}\alpha_i) = \widetilde{\tau}\big((\tau\alpha)_{\theta\alpha(i)}\big) = \big(\tau(\tau\alpha)\big)_{\theta\tau\alpha(\theta\alpha(i))} = \alpha_i,$$

which implies that $\widetilde{\tau}$ is an involution. Furthermore, the triple point relations in $\pi(\mathcal{D})$ imply that $\widetilde{\tau}$ groups the crossings of $\widetilde{\mathcal{D}}$ in triplets of pairwise related points. Accordingly, $(\widetilde{\mathcal{D}}, \widetilde{\tau})$ is a diagram in $\mathbb{S}^2 \times \mathbf{n}$.

The diagram $\widetilde{\mathcal{D}}$ defines as usual an equivalence relation '$\sim_{\widetilde{\mathcal{D}}}$' in the surface $\mathbb{S}^2 \times \mathbf{n}$. Denote by $\widetilde{\Sigma}$ the quotient space of $\mathbb{S}^2 \times \mathbf{n}$ by this relation. Let $\widetilde{f} : \mathbb{S}^2 \times \mathbf{n} \to \widetilde{\Sigma}$ be the associated quotient map.

There exists a unique map $p : \widetilde{\Sigma} \to \Sigma$ making the following diagram commutative:

$$
\begin{array}{ccc}
\mathbb{S}^2 \times \mathbf{n} & \xrightarrow{\;\widetilde{f}\;} & \widetilde{\Sigma} \\
{\scriptstyle \widetilde{p}}\Big\downarrow & & \Big\downarrow{\scriptstyle p} \\
\mathbb{S}^2 & \xrightarrow[\;f\;]{} & \Sigma
\end{array}
$$

where $\widetilde{p}$ is the first coordinate projection. Any point $\widetilde{Y} \in \widetilde{\Sigma}$ has the form $\widetilde{Y} = \widetilde{f}(Y, i)$, for some $Y \in \mathbb{S}^2$ and $i \in \mathbf{n}$. Thus, we define $p(\widetilde{Y}) = f(Y)$. By construction, $\widetilde{p}$ maps points of $\mathbb{S}^2 \times \mathbf{n}$ related by $\widetilde{\mathcal{D}}$ into points of $\mathbb{S}^2$ related by $\mathcal{D}$, therefore p is well defined. It is easy to check that p is an n-sheeted covering map of Σ, in particular it is continuous.

By construction, it is straightforward to see that θ is the representation associated to p.

If $\Sigma_i = \widetilde{f}(\mathbb{S}^2 \times \{i\})$ for $i \in \mathbf{n}$, then

$$
\widetilde{\Sigma} = \bigcup_{i \in \mathbf{n}} \Sigma_i,
$$

and for $i, j \in \mathbf{n}$, the two 2-spheres Σ_i and Σ_j meet if and only if there exists a curve $\alpha \in \mathcal{D}$ with $\theta\alpha(i) = j$. If we denote by $\Omega(\mathcal{D})$ the image subgroup of θ in Ω_n, Σ_i and Σ_j are in the same connected component of $\widetilde{\Sigma}$ if and only if there exists $\omega \in \Omega(\mathcal{D})$ with $\omega(i) = j$. Then it follows that $\widetilde{\Sigma}$ is connected if and only if θ is transitive. $\qquad\Box$

We will construct a generating system for $\pi_1(\Sigma)$ associated to the diagram $\mathcal{D}$ that verifies analogous relations as those of the diagram group.

Fix the simple point X_0 as base point of the fundamental group of Σ. Let $X \in \mathbb{S}^2$ be the preimage of X_0 under f. We write $a \simeq b$ to denote that the paths a and b (in the space they belong to) are homotopic relative to their endpoints. If a is a path in $\mathbb{S}^2$, $\overline{a}$ denotes the path $f \circ a$ in Σ.

Consider two sister curves $\alpha, \tau\alpha \in \mathcal{D}$ and points $A, \tau A \in \mathbb{S}^2$ related by $\mathcal{D}$ such that $A \in |\alpha|$ and $\tau A \in |\tau\alpha|$. Let $\lambda_\alpha : I \to \mathbb{S}^2$ be a path in $\mathbb{S}^2$ from X to A. Similarly, consider a path $\lambda_{\tau\alpha} : I \to \mathbb{S}^2$ from X to τA. Since A and τA are related by $\mathcal{D}$, both paths $\overline{\lambda}_\alpha$ and $\overline{\lambda}_{\tau\alpha}$ have the same endpoints, therefore $\overline{\lambda}_\alpha * \overline{\lambda}_{\tau\alpha}^{-1}$ is a closed path in Σ based at X_0. The *dual loop* $\alpha_{\#}$ of α is the class in $\pi_1(\Sigma, X_0)$ of the path $\overline{\lambda}_\alpha * \overline{\lambda}_{\tau\alpha}^{-1}$.

Lemma 4.5. *The definition of $\alpha_{\#}$ does not depend on the election of A and τA nor on the election of the paths λ_α and $\lambda_{\tau\alpha}$.*

Proof. If λ'_α and $\lambda'_{\tau\alpha}$ are paths with the same endpoints points as λ_α and $\lambda_{\tau\alpha}$ respectively, then $\lambda_\alpha \simeq \lambda'_\alpha$ and $\lambda_{\tau\alpha} \simeq \lambda'_{\tau\alpha}$ because they are paths in $\mathbb{S}^2$. Therefore $\overline{\lambda}_\alpha * \overline{\lambda}_{\tau\alpha} \simeq \overline{\lambda}'_\alpha * \overline{\lambda}'_{\tau\alpha}$.

Consider two related points $A' \in |\alpha|$ and $\tau A' \in |\tau\alpha|$ different from A and τA, and two paths λ''_α and $\lambda''_{\tau\alpha}$ connecting X with A and τA, respectively.

Let $a : I \to |\alpha|$ be a path with $a(0) = A'$ and $a(1) = A$. We can assume that a is an immersion, and so that it follows the course of α at the self-intersection points of α. In this conditions, the path a has a "sister" path $\tau a : I \to |\tau\alpha|$ with $\tau a(0) = \tau A'$ and $\tau a(1) = \tau A$, such that $a(t)$ and $\tau a(t)$ are related by $\mathcal{D}$ for all $t \in I$.

The paths $\lambda''_\alpha * a$ and $\lambda''_{\tau\alpha} * \tau a$ in $\mathbb{S}^2$ have the same endpoints as λ_α and $\lambda_{\tau\alpha}$, respectively. As before, this implies that the closed paths $\overline{\lambda}_\alpha * \overline{\lambda}_{\tau\alpha}^{-1}$ and

$$\overline{(\lambda''_\alpha * a)} * \overline{(\lambda''_{\tau\alpha} * \tau a)}^{-1} = \overline{\lambda}''_\alpha * \overline{a} * \overline{\tau a}^{-1} * \overline{\lambda}''^{-1}_{\tau\alpha} \simeq \overline{\lambda}''_\alpha * \overline{\lambda}''^{-1}_{\tau\alpha}$$

define the same element of $\pi_1(\Sigma, X_0)$. □

Let $\mathcal{D}_\#$ denote the set of dual loops of the curves of $\mathcal{D}$.

Proposition 4.6. *The set $\mathcal{D}_\#$ generates $\pi_1(\Sigma, X_0)$.*

Proof. Let $\overline{z} : I \to \Sigma$ be a closed path in Σ based at X_0.

Up to homotopy, we can assume that $\overline{z}$ meets no triple point of Σ, and that there is a finite number k of points of I whose image under $\overline{z}$ is a double point of Σ. In this case, if $\overline{z}(t)$ is a double point of Σ for some $t \in I$, there are two cases: (i) there is a sufficiently small $\varepsilon > 0$ such that $\overline{z}([t - \varepsilon, t + \varepsilon])$ is completely contained in one of the two sheets of Σ intersecting at $\overline{z}(t)$; or (ii) there is a sufficiently small $\varepsilon > 0$ such that $\overline{z}([t - \varepsilon, t])$ and $\overline{z}([t, t + \varepsilon])$ are contained in different sheets of Σ intersecting at $\overline{z}(t)$. In the latter case we say that the path $\overline{z}$ *changes of sheet at t*.

Let $t_1 < \cdots < t_m$ be the points of I at which $\overline{z}$ changes of sheet, and take also $t_0 = 0, t_{m+1} = 1$. For each $j = 0, \ldots, m$ there exists a path $z_i : [t_i, t_{i+1}] \to \mathbb{S}^2$ such that $\overline{z}_i = \overline{z}|_{[t_i, t_{i+1}]}$. Then, $\overline{z} = \overline{z}_0 * \overline{z}_1 * \cdots * \overline{z}_m$.

If $m = 0$ or 1, either $\overline{z}$ is homotopically trivial or it is a representative of an element of $\mathcal{D}_\#$, and so there is nothing to prove.

Assume that $m \geq 2$. For each $j = 1, \ldots, m - 1$, the path z_j can be homotopically deformed in $\mathbb{S}^2$ (fixing the endpoints) into a path z'_j passing through X. This path z'_j can be written as a product $(\lambda'_j)^{-1} * \lambda_{j+1}$, where λ'_j is a path in $\mathbb{S}^2$ from X to $z_j(t_j)$, and λ_{j+1} is a path in $\mathbb{S}^2$ from X to $z_j(t_{j+1})$. If we define $\lambda_1 = z_0$ and $\lambda'_m = z_m^{-1}$, then

$$\overline{z} \simeq \overline{\lambda}_1 * \overline{\lambda'}_1^{-1} * \overline{\lambda}_2 * \overline{\lambda'}_2^{-1} * \cdots * \overline{\lambda}_{m-1} * \overline{\lambda'}_{m-1}^{-1} * \overline{\lambda}_m * \overline{\lambda'}_m^{-1},$$

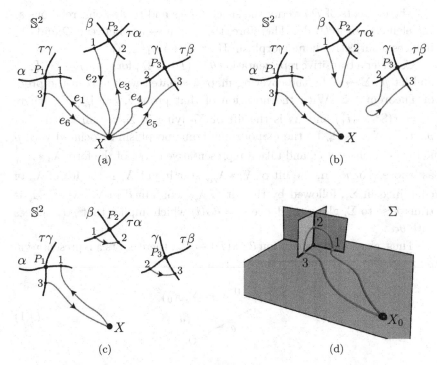

Fig. 4.2: Triple point relation for dual loops.

where the right-hand side of the expression is a product of representatives of elements of $\mathcal{D}_\#$. $\qquad\square$

Proof of Theorem 4.3. If $\alpha \in \mathcal{D}$ and we choose a representative $\overline{\lambda}_\alpha * \overline{\lambda}_{\tau\alpha}^{-1}$ of its dual loop $\alpha_\#$, then its inverse path $\overline{\lambda}_{\tau\alpha} * \overline{\lambda}_\alpha^{-1}$ is a representative of $(\tau\alpha)_\#$. In particular, $(\tau\alpha)_\# = \alpha_\#^{-1}$ for all $\alpha \in \mathcal{D}$.

On the other hand, if P_1, P_2, P_3 form a triplet of $\mathcal{D}$, if the curves meeting at them are labeled as in the standard triple point configuration, a representative of $\alpha_\# \beta_\# \gamma_\# \in \pi_1(\Sigma, X_0)$ is

$$\overline{e}_1 * \overline{e}_2 * \overline{e}_3 * \overline{e}_4 * \overline{e}_5 * \overline{e}_6,$$

for some paths $e_1, \ldots, e_6$ of $\mathbb{S}^2$ as in Fig. 4.2(a). Figure 4.2 shows how this product of paths can be homotopically deformed into the situation of Fig. 4.2(c), which corresponds to the contractible path in Σ depicted in Fig. 4.2(d). Hence,

$$\alpha_\# \beta_\# \gamma_\# = 1.$$

The elements of $\mathcal{D}_{\#}$ verify the sister curves and triple point relations as the elements of $\pi(\mathcal{D})$ do. Therefore, the map $\alpha \mapsto \alpha_{\#}$ between $\mathcal{D}$ and $\mathcal{D}_{\#}$ induces a surjective homomorphism $\Pi : \pi(\mathcal{D}) \to \pi_1(\Sigma, X_0)$.

Consider a transitive representation $\theta : \pi(\mathcal{D}) \to \Omega_n$ for some $n \in \mathbb{N} \cup \{\infty\}$, and let $p : \widetilde{\Sigma} \to \Sigma$ be the covering map associated with θ as in the proof of Theorem 4.4. With the notation of that proof $\widetilde{\Sigma} = \bigcup_{i \in \mathbf{n}} \Sigma_i$, where $\Sigma_i = \widetilde{f}(\mathbb{S}^2 \times \{i\})$ and X_i is the lift of X_0 lying in Σ_i for all $i \in \mathbf{n}$. Let $\rho : \pi_1(\Sigma, X_0) \to \Omega_n$ be the monodromy homomorphism associated with p. Let α be a curve of $\mathcal{D}$, and take a representative of $\alpha_{\#}$ of the form $\overline{\lambda}_\alpha * \overline{\lambda}_{\tau\alpha}^{-1}$ as before. For $i \in \mathbf{n}$, the lift of $\overline{\lambda}_\alpha * \overline{\lambda}_{\tau\alpha}^{-1}$ starting at X_i is the lift of $\overline{\lambda}_\alpha$ of contained in Σ_i, followed by the lift of $\overline{\lambda}_{\tau\alpha}^{-1}$ contained in Σ_j, where Σ_j is transverse to Σ_i along α. Hence, $j = \theta\alpha(i)$ which implies that $\rho\alpha_{\#}$ agrees with $\theta\alpha$.

Thus, given a representation $\theta : \pi(\mathcal{D}) \to \Omega_n$, there exists a representation $\rho : \pi_1(\Sigma, X_0) \to \Omega_n$ making the diagram

$$\pi(\mathcal{D}) \xrightarrow{\ \Pi\ } \pi_1(\Sigma, X_0)$$

(4.1)

commutative.

Due to the Cayley Theorem [25, p. 9], for any group G there exists a faithful representation of G into the group of permutations of its own elements. Since $\pi(\mathcal{D})$ is finitely generated, it is countable, and the group of permutations of its elements is isomorphic to the group Ω_n for some $n \in \mathbb{N} \cup \{\infty\}$. Hence, there always exists a faithful representation $\theta : \pi(\mathcal{D}) \to \Omega_n$. If ρ is taken as in Eq. (4.1), both ρ and Π are injective. Since Π is onto, it is an isomorphism. $\qquad\square$

4.4   Applications

If Σ is a filling Dehn sphere of a 3-manifold M, the fundamental group of M is isomorphic to the fundamental group of Σ. So it is possible to obtain a presentation of the fundamental group of a 3-manifold in terms of a Johansson representation of the 3-manifold in $\mathbb{S}^2$.

Assume that in a triplet of a diagram $\mathcal{D}$ there is one crossing which is a self-intersection point of a diagram curve. For example, suppose that in Fig. 3.1 the curves α and $\tau\gamma$ agree. Then, $\gamma = \tau\alpha$, and the triple point relation

$$\alpha\beta\gamma = 1$$

becomes

$$\alpha\beta\alpha^{-1} = 1,$$

or, equivalently, $\beta = 1$. Following [23], in this case we say that the pair of sister curves β, $\tau\beta$ of $\mathcal{D}$ is *compensated*. It trivially follows that a diagram in which every pair of sister curves is compensated is simply connected, i.e., the corresponding Dehn sphere or pseudo Dehn sphere is simply connected. A related result is the following unpublished one due to Haken:

Theorem 4.7. *Let $\mathcal{D}$ be a connected diagram in $\mathbb{S}^2$ with only two curves. There are two possibilities: the two curves of $\mathcal{D}$ are simple and $\pi(\mathcal{D}) \simeq \mathbb{Z}_3$; or both curves are not simple and $\pi(\mathcal{D}) \simeq 1$.*

Proof. Denote by α and $\tau\alpha$ the two curves of $\mathcal{D}$. Counting crossings, it follows that α is simple if and only if $\tau\alpha$ is simple.

If both are non simple curves, they are compensated and then $\pi(\mathcal{D}) \simeq 1$.

If both curves are simple, all the triple point relations are of the form $\alpha\alpha\alpha = 1$ (in Fig. 3.1 this case corresponds to $\alpha = \beta = \gamma$), which clearly implies $\pi(\mathcal{D}) \simeq \mathbb{Z}_3$. □

The Dehn spheres (or pseudo Dehn spheres) given by the diagrams of Figs. 3.2(a) and 3.2(b) are simply connected, those of Figs. 3.2(c), 3.2(e) and 3.2(f) verify $\pi(\mathcal{D}) = \mathbb{Z}$, and for the diagram in Fig. 3.2(d) $\pi(\mathcal{D}) = \mathbb{Z}_3$.

Recall from Sec. 3.4 that a connected diagram $\mathcal{D}$ in $\mathbb{S}^2$ is filling if and only if the boundary of $\widehat{M}(\mathcal{D})$ is a disjoint union of 2-spheres. If the boundary of $\widehat{M}(\mathcal{D})$ has a connected component not homeomorphic to a 2-sphere, then $H_1(\partial M, \mathbb{Z}_2) \neq 0$. Using the long exact sequence with coefficients in $\mathbb{Z}_2$ associated to $(M, \partial M)$ it follows that $H_1(M, \mathbb{Z}_2) \neq 0$, and this implies that $\pi_1(\widehat{M})$ has an index 2 subgroup (see [27, Lemma 4.9]). Therefore:

Theorem 4.8. *If $\mathcal{D}$ is a realizable connected diagram in $\mathbb{S}^2$ such that $\pi(\mathcal{D})$ has no index 2 subgroup, then $\mathcal{D}$ is filling.* □

Corollary 4.9. *If $\mathcal{D}$ is a realizable connected diagram in $\mathbb{S}^2$ with only two curves, it is filling.*

Proof. By Theorem 4.7, if $\mathcal{D}$ is connected and it only has two curves, then the diagram group is trivial or cyclic of order 3. In any case, there is no index 2 subgroups. By Theorem 4.8, $\mathcal{D}$ is filling. □

This result, together with Theorem 4.7, bring us to the following conjecture of Haken which, like a plenty of other problems, became solved after Perelman's proof of the Thurston Geometrization Conjecture [52–54].

Conjecture 4.10. *If $\mathcal{D}$ is a realizable connected diagram in $\mathbb{S}^2$ with only two curves, then $M(\mathcal{D})$ is homeomorphic to $\mathbb{S}^3$ or to the lens space $L(3,1)$.*

In particular, the diagram in Fig. 3.2(a) and the diagram in Fig. 3.2(d) are filling. As expected, the first of them is a Johansson representation of $\mathbb{S}^3$ (see Sec. 7.1) and the second one represents the lens space $L(3,1)$ (see Sec. 7.2).

Remark 4.11. The presentation of the fundamental group of a Dehn sphere given by Theorem 4.3 is also valid for Dehn disks.

4.5   The fundamental group of a Dehn g-torus

The presentation of the fundamental group of a Dehn sphere given by Theorem 4.3 can be extended to a generic Dehn surface with connected domain. We show here, without proofs, the main points of that construction. For details see [42].

Let $\mathcal{D}$ be a diagram in a connected and orientable surface S, which could have nonempty boundary. As in the previous case, Σ is the quotient space of S under the relation induced by $\mathcal{D}$, and $f : S \to \Sigma$ is the canonical projection.

A path in Σ is *surfacewise* if it is a path on S mapped to Σ through f. Take a simple point $X_0 \in \Sigma$ as basepoint of $\pi_1(\Sigma)$. Let $\pi_S = f_*(\pi_1(S))$ be the subgroup of $\pi_1(\Sigma)$ generated by the surfacewise loops based at X_0. A dual loop to a curve $\alpha \in \mathcal{D}$ can be defined as the class in $\pi_1(\Sigma, X_0)$ of a closed path of the form $\overline{\lambda}_\alpha * \overline{\lambda}_{\tau\alpha}^{-1}$, where $\overline{\lambda}_\alpha, \overline{\lambda}_{\tau\alpha}$ are constructed exactly as it is done in the case $S = \mathbb{S}^2$. It this case, since S is not assumed to be simply connected, the proof of Lemma 4.5 breaks down and dual loops to α are not unique. However:

Lemma 4.12. *Two loops $\alpha_\#$ and $\alpha'_\#$ on Σ dual to $\alpha \in \mathcal{D}$ are surfacewise conjugate, that is, there are two surfacewise loops s and t based at X_0 such that $\alpha_\# = s\alpha'_\# t$.*   □

Proposition 4.13. *Surfacewise loops based at X_0 and loops dual to the curves of $\mathcal{D}$ generate $\pi_1(\Sigma)$.*   □

The proof of this result is similar to that of Proposition 4.6.

Since π_S is finitely generated and $\mathcal{D}$ is finite, we can choose a finite set of generators of $\pi_1(\Sigma)$ by taking a finite generating set $\mathcal{B}_S = \{s_1, s_2, \ldots, s_k\}$ of π_S and a set $\mathcal{D}_\#$ that contains exactly one dual loop for each curve of $\mathcal{D}$. The elements of $\mathcal{D}_\#$ are called *preferred dual loops* of the curves of $\mathcal{D}$.

In order to obtain a presentation of $\pi_1(\Sigma)$, a set of relators associated to these generators must be established. These relators will be a natural extension of the relators for Dehn spheres:

(R1) *Sister curve relations.* By construction, if $\alpha, \tau\alpha$ are two sister curves of $\mathcal{D}$, their dual loops $\alpha_\#, (\tau\alpha)_\#$ can be chosen verifying $(\tau\alpha)_\# = \alpha_\#^{-1}$.

(R2) *Triple point relations.* Let P be a triple point of Σ and let P_1, P_2, P_3 be the three crossings of $\mathcal{D}$ in the triplet of P. If we label the curves of $\mathcal{D}$ intersecting at these crossings as in Figure 3.1, we get the triple point relation around P:

$$t_1 \alpha_\# t_2 \beta_\# t_3 \gamma_\# t_4 = 1, \qquad (4.2)$$

where $t_1, \ldots, t_4$ are surfacewise loops that can be expressed explicitly as words in $s_1, \ldots, s_k$.

(R3) *Double curve relations.* This relation arises because when we identify sister curves of $\mathcal{D}$ in Σ, different elements of $\pi_1(S)$ become identified in Σ. Given $\alpha \in \mathcal{D}$, let λ_α and $\lambda_{\tau\alpha}$ be the paths from X_0 to $|\alpha|$ and $|\tau\alpha|$, respectively, used for constructing the preferred dual loop $\alpha_\#$. Consider now the curves $\alpha, \tau\alpha$ as closed paths starting at $\lambda_\alpha(1)$ and $\lambda_{\tau\alpha}(1)$, respectively, and denote again by α and $\tau\alpha$ the surfacewise elements of $\pi_1(\Sigma)$ represented by $\overline{\lambda_\alpha * \alpha * \lambda_\alpha^{-1}}$ and $\overline{\lambda_{\tau\alpha} * \tau\alpha * \lambda_{\tau\alpha}^{-1}}$, respectively. This relation can be written as

$$\alpha\, \alpha_\# = \alpha_\# \tau\alpha. \qquad (4.3)$$

(R4) *Surface relations.* Those are the relations that the surfacewise generators $s_1, s_2, \ldots, s_{2g}$ verify when considered as elements of $\pi_1(S)$.

Theorem 4.14 ([42]). *The fundamental group $\pi_1(\Sigma)$ is isomorphic to the one generated by $\mathcal{B}_S \cup \mathcal{D}_\#$ together with the above relations.* $\square$

Example 4.15. Let $\mathcal{D}$ be the diagram on the torus $\mathbb{T}^2$ depicted in Fig. 3.2(h). Let Σ be the quotient space of $\mathbb{T}^2$ under the relation induced by $\mathcal{D}$. Let α and $\tau\alpha$ be the two curves of $\mathcal{D}$, and take the points $X, A, \tau A$ and the paths $\lambda_\alpha, \lambda_{\tau\alpha}$ as in Fig. 4.3. In the same figure, we have depicted a pair of generators a, b of the fundamental group of $\mathbb{T}^2$ based at

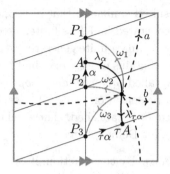

Fig. 4.3: The group $\pi(\mathcal{D})$ in an example on $\mathbb{T}^2$.

X and three paths $\omega_1, \omega_2, \omega_3$ connecting X with the crossings P_1, P_2, P_3 respectively. For $i = 1, 2, 3$, let d_i be a path in $|\alpha|$ that connects P_i with A following the course of α, and let τd_i be its sister path in $|\tau \alpha|$.

For simplicity, objects in $\mathbb{T}^2$ are denoted as their images in Σ. Take $\mathcal{B}_{\mathbb{T}^2} = \{a, b\}$, and consider the paths $\alpha_\# = \lambda_\alpha * \lambda_{\tau\alpha}^{-1}$ and $(\tau \alpha)_\# = \lambda_{\tau\alpha} * \lambda_\alpha^{-1}$. It is clear that the sister curve and surface relations given by $\mathcal{D}$ are $(\tau\alpha)_\# = \alpha_\#^{-1}$ and $aba^{-1}b^{-1} = 1$.

Triple point relation. The loop
$$\omega_1 * \omega_2^{-1} * \omega_2 * \omega_3^{-1} * \omega_3 * \omega_1^{-1}$$
is homotopically trivial in Σ. On the other hand, this path is homotopic relative to its extremes to the path
$$\omega_1 d_1 \lambda_\alpha^{-1} \, \lambda_\alpha \lambda_{\tau\alpha}^{-1} \, \lambda_{\tau\alpha} \tau d_1^{-1} \omega_2^{-1}$$
$$\omega_2 d_2 \lambda_\alpha^{-1} \, \lambda_\alpha \lambda_{\tau\alpha}^{-1} \, \lambda_{\tau\alpha} \, \tau d_2^{-1} \omega_3^{-1}$$
$$\omega_3 d_3 \lambda_\alpha^{-1} \, \lambda_\alpha \lambda_{\tau\alpha}^{-1} \, \lambda_{\tau\alpha} \, \tau d_3^{-1} \omega_1^{-1},$$
where the stars have been dropped from notation for simplicity. Hence,
$$a \, \alpha_\# \, a^{-1} b^{-2} \, \alpha_\# \alpha_\# \, a^{-1} b^{-1} = 1.$$
Double curve relation. For convenience, denote the closed paths in $\mathbb{T}^2$ that travel along $|\alpha|$ and $|\tau\alpha|$ starting and finishing at A and τA by α and $\tau\alpha$, respectively. In Σ,
$$\lambda_\alpha * \alpha * \lambda_\alpha^{-1} \simeq \lambda_\alpha * \lambda_{\tau\alpha}^{-1} * \lambda_{\tau\alpha} * \tau\alpha * \lambda_{\tau\alpha}^{-1} * \lambda_{\tau\alpha} * \lambda_\alpha^{-1}.$$
Therefore,
$$a = \alpha_\# ab^3 (\tau\alpha)_\#.$$
In summary, a presentation of the fundamental group of Σ is
$$\langle \, a, b, \alpha_\# \mid a\alpha_\# a^{-1} b^{-2} \alpha_\#^2 a^{-1} b^{-1}, \; \alpha_\# ab^3 \alpha_\#^{-1} a^{-1}, \; aba^{-1}b^{-1} \, \rangle.$$

Chapter 5

Filling homotopies

5.1 Filling homotopies

Let S be a surface and let M be a 3-manifold.

If $f, g : S \to M$ are regularly homotopic immersions, they are indeed homotopic, but the converse is not always true: there are homotopic immersions which are not regularly homotopic.

Theorem 5.1 (Thm. 1.1 in [26], cf. Thm. 6 [41]). *The regular homotopy classes within the homotopy class of an immersion $f : S \to M$ are in one to one correspondence with $H^1(S; \mathbb{Z}_2)$, unless both S and M are non-orientable, in this case the correspondence is two to one.* □

Since $H^1(\mathbb{S}^2; \mathbb{Z}_2)$ is trivial:

Corollary 5.2. *Two immersions of the 2-sphere into a 3-manifold are regularly homotopic if and only if they are homotopic.* □

In particular, two parametrizations of null-homotopic filling Dehn spheres in M are regularly homotopic. Recall that in an irreducible 3-manifold every Dehn sphere is null-homotopic.

In [32], Homma and Nagase introduced a set of elementary deformations or *moves* for immersions of surfaces into 3-manifolds. Following [32], these *Homma-Nagase moves* are the elementary deformations of types I, II, III and VI depicted in Figs. 5.1 to 5.4.

Theorem 5.3 ([33]). *Two transverse immersions of the same surface S into a 3-manifold M are regularly homotopic if and only if it is possible to transform them into each other by a finite sequence of Homma-Nagase moves and ambient isotopies of M.* □

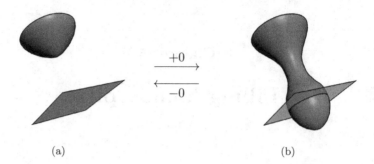

(a)   (b)

Fig. 5.1: Homma-Nagase move of type I or finger move 0.

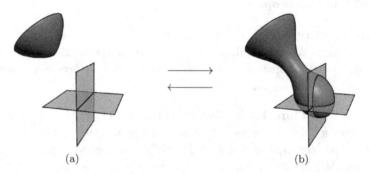

(a)   (b)

Fig. 5.2: Homma-Nagase move of type II.

This theorem is stated in [33] in the PL category. A proof in the differentiable category is indicated in [57]. See [26, Theorem 3.1] for an equivalent result, also in the differentiable category.

The *Haken moves*[1] are obtained replacing the type II and III Homma-Nagase moves with the finger moves 1 and 2 depicted in Figs. 5.5 and 5.6 respectively. In this notation, the Homma-Nagase moves of type I and VI are called *finger move* 0 and *saddle move* respectively. A finger move i, with $i = 0$, 1 or 2, is a finger move $+i$ if the deformation occurs from left to right in the corresponding figure. Otherwise it is a finger move $-i$. A saddle move is "self-symmetric", i.e., it is equivalent in both directions: the saddle move from left to right in Fig. 5.4 can be seen as a "push" of a sheet of Σ along a 2-gon (see Figs. 5.9 and 5.11(b), and Sec. 6.3.3), that can be undone by a similar push along a different 2-gon (see Fig. 5.11(b)).

---

[1] All these moves, except finger move 2, were introduced in [23].

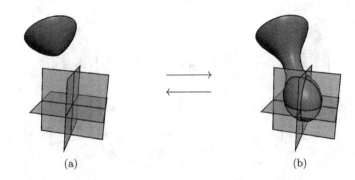

(a) (b)

Fig. 5.3: Homma-Nagase move of type III.

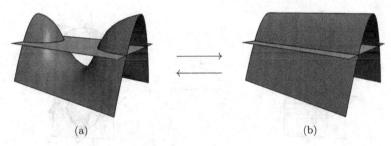

(a) (b)

Fig. 5.4: Homma-Nagase move of type VI or saddle move.

Lemma 5.4. *The Homma-Nagase and Haken sets of moves are equivalent up to ambient isotopy.*

Proof. It is enough to show that each Homma-Nagase move can be decomposed into a finite sequence of Haken moves (and ambient isotopies) and vice versa. Figures 5.7(a) to 5.7(d) shows how a type II Homma-Nagase move can be decomposed into a finger move +0 (Fig. 5.7(b)), followed by a finger move +1 (Fig. 5.7(c)), with a final ambient isotopy (Fig. 5.7(d)). Adding a finger move +2 (Fig. 5.7(e)), we obtain a move equivalent to a type III Homma-Nagase move (Fig. 5.7(f)). The equivalence in the other direction can be shown in a similar way. □

Thus, it is possible to replace the Homma-Nagase moves with the Haken moves in the statement of Theorem 5.3. The latter are better suited to the study of filling surfaces.

Definition 5.5. A Haken move performed on a filling immersion is a *filling move*, or simply an *f-move*, if the immersion obtained after it is also filling.

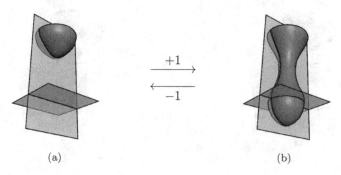

Fig. 5.5: Finger move 1.

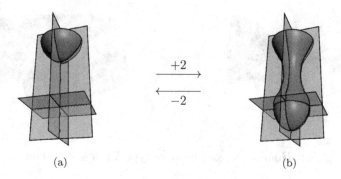

Fig. 5.6: Finger move 2.

Lemma 5.6. *For Haken moves applied to filling immersions, we have:*

*(1) finger moves 0 never are **f**-moves;*
*(2) finger moves +1 or ±2 always are **f**-moves; and*
*(3) finger moves −1 and saddle moves might be **f**-moves or not.*   □

This lemma is straightforward from Proposition 2.7 and Figs. 5.1, 5.4 , 5.5 and 5.6.

Definition 5.7. Two filling immersions $f, g : S \to M$ are *filling homotopic*, or simply ***f**-homotopic*, if there exists a finite sequence of immersions

$$f = f_0, f_1, \dots, f_n = g$$

such that f_i and f_{i+1} are either ambient isotopic or related by an f-move, for all $i = 0, \dots, n - 1$.

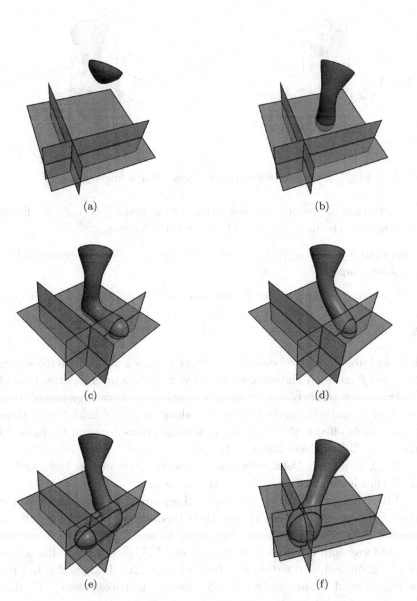

(a)

(b)

(c)

(d)

(e)

(f)

Fig. 5.7: Homma-Nagase moves in terms of Haken moves.

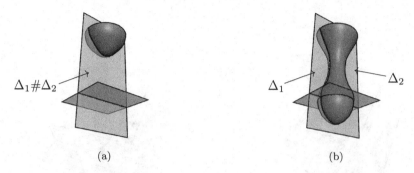

Fig. 5.8: Modifications on the faces after a finger move 1.

Note that in the previous definition all the maps $f_0, \ldots, f_n$ are filling immersions. The main result of the book is the following:

Theorem 5.8. *If $f, g : \mathbb{S}^2 \to M$ are null-homotopic filling immersions, they are $\boldsymbol{f}$-homotopic.*

The proof of Theorem 5.8 is quite long, so it is postponed until Chap. 6.

5.2   Bad Haken moves

It is desirable to understand when a finger move -1 or a saddle move fails to be an $\boldsymbol{f}$-move. A Haken move on a Dehn surface produces two kind of effects on the i-cells of Σ ($i \geq 1$): new connections between some of these i-cells of Σ, and some i-cells of Σ get cut along an $(i-1)$-disk. Among these two possible effects, if Σ is filling, new connections between i-cells could transform Σ into a non-filling surface.

Let Σ be a filling Dehn surface in M, and let Σ' be another Dehn surface in M that is obtained from Σ by a Haken move.

Under a finger move -1, two trihedral regions of Σ disappear while the rest of regions of Σ do not change their topological type. Hence, all the regions of Σ' are open 3-balls. Two faces Δ_1 and Δ_2 of Σ get connected by the move into a unique face $\Delta_1 \# \Delta_2$ of Σ' (Fig. 5.8). If Σ' has a face not homeomorphic to a disk, it is because $\Delta_1 = \Delta_2$. If $\Delta_1 \neq \Delta_2$, but the double curve of Σ' surrounding $\Delta_1 \# \Delta_2$ contains no triple points of Σ', then both faces Δ_1 and Δ_2 should be 1-gons. Therefore:

Proposition 5.9. *With the notation of Fig. 5.8, a finger move -1 applied on a filling Dehn surface is not an $\boldsymbol{f}$-move if and only if $\Delta_1 = \Delta_2$, or $\Delta_1 \neq \Delta_2$ and both Δ_1 and Δ_2 are 1-gons.* $\qquad\square$

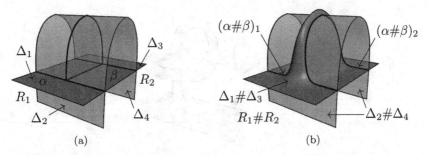

Fig. 5.9: Modifications on the faces after a saddle move.

Assume now that Σ' is obtained from Σ after the saddle move depicted in Fig. 5.9. This move connects two regions R_1 and R_2 of Σ into a unique region $R_1 \# R_2$ of Σ', and a third region R_3 of Σ gets split along a disk. If a region of Σ' is not a 3-ball, such region must be[2] $R_1 \# R_2$.

Within the faces of Σ, there are four faces (called Δ_1, Δ_2, Δ_3 and Δ_4 in Fig. 5.9(a)), which get pairwise connected by the move, producing two faces $\Delta_1 \# \Delta_3$ and $\Delta_2 \# \Delta_4$ of Σ'. In addition, there are two faces Δ_5, Δ_6 of Σ each of which become split into two faces of Σ'. A face of Σ' that is not an open 2-disk must be necessarily $\Delta_1 \# \Delta_3$ or $\Delta_2 \# \Delta_4$.

At the double curves of Σ, the saddle move connects two edges α and β of Σ into another two edges $(\alpha \# \beta)_1$ and $(\alpha \# \beta)_2$ of Σ' (see Fig. 5.9). If Σ' has a double curve without triple points, it must be because one of the edges $(\alpha \# \beta)_1$ or $(\alpha \# \beta)_2$ closes without meeting any triple point, which is equivalent to $\alpha = \beta$.

Therefore:

Proposition 5.10. *With the notation of Fig. 5.9, a saddle move applied on a filling Dehn surface is not an **f**-move if and only if one of the following conditions holds:*

(1) $R_1 \# R_2$ is not an open 3-ball;
(2) $\Delta_1 \# \Delta_3$ or $\Delta_2 \# \Delta_4$ is not an open 2-disk; and
(3) $\alpha = \beta$. □

From this proposition, a particular case in which it is possible to ensure that the saddle move is an **f**-move is the following.

---

[2]In the extreme case where the three regions R_1, R_2 and R_3 are the same, it could happen that the connection between R_1 and R_2 got 'neutralized' by the splitting in R_3, so the resulting region is a 3-ball.

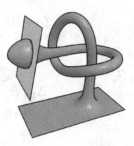

Fig. 5.10: A knotted hole after a finger move +0.

Proposition 5.11. *The saddle move of Fig. 5.9 applied to a filling Dehn surface is an f-move if it fulfills the following three conditions:*

(1) $R_1 \neq R_2$;
(2) $\Delta_i \neq \Delta_j$, for all $i, j \in \{1, 2, 3, 4\}, i \neq j$; and
(3) $\alpha \neq \beta$.   □

5.3   "Not so bad" Haken moves

In this section we want to remark that, among the bad Haken moves, some of them are worse than others.

Let Σ be a Dehn surface of M, and let Σ' be a Dehn surface of M obtained from Σ after a Haken move.

Assume that the move transforming Σ into Σ' is a finger move +0. Let R be the region of Σ in which the finger move takes place and let R' be the region of Σ' coming from R after the move. The region R' is obtained from R after removing a cylinder $\mathbb{D}^2 \times I$ embedded in $\mathrm{cl}(R)$ with $\mathbb{D}^2 \times I \cap \partial R = \mathbb{D}^2 \times \partial I$. For example, if R is an open 3-disk, then R' can be an open solid torus, or an open 3-ball with a knotted hole (Fig. 5.10). If we look at the inverse finger move -0, $\mathrm{cl}(R)$ is obtained from $\mathrm{cl}(R')$ after attaching a 2-handle. Thus, in general there is little control on the topological type of the regions that result after a finger move ± 0. Hence, finger moves 0 are the "worst" Haken moves.

If Σ' is obtained form Σ after a finger move 1 or 2, the topological type of a region of Σ' coincides with that of a region of Σ.

Finally, assume that Σ' is obtained from Σ after a saddle move. With the notation of Fig. 5.9, the discussion of Sec. 5.2 implies that the saddle move has the following possible effects on the regions of Σ: if $R_1 \neq R_2$, then we are making the connected sum of these two regions; if $R_1 = R_2$, then we are adding a 1-handle to this region; and finally, we are removing a 1-handle to R_3. In particular, if the 3-disks are regarded as handlebodies of genus $g = 0$, if all the regions of Σ are open handlebodies, all the regions of Σ' are open handlebodies.

Proposition 5.12. *If the move relating Σ and Σ' is not a finger move 0 and all the regions of Σ are open handlebodies, then all the regions of Σ' are open handlebodies.* $\square$

Definition 5.13. A Dehn surface is *eventually filling* if its Johansson diagram is filling.

If Σ is eventually filling in M, it might not fill M, but there exists a 3-manifold M' with a filling Dehn surface Σ' homeomorphic to Σ. For instance, if we consider any filling Dehn surface Σ in $\mathbb{S}^3$, we can think of Σ as contained in a closed 3-disk $B^3 \subset \mathbb{S}^3$. If M is another 3-manifold and $e : B^3 \to M$ is an embedding, the Dehn surface $e(\Sigma)$ is eventually filling.

Definition 5.14. Two immersions $f, g : S \to M$ are *weakly f-homotopic* if they can be transformed into each other by a finite sequence of saddle moves, finger moves ± 1 and ± 2, and ambient isotopies.

Theorem 5.15. *Let $f : S \to M$ be a filling immersion and let $g : S \to M$ be an eventually filling immersion. If f and g are weakly f-homotopic, then g is filling.*

Proof. Since g is weakly homotopic to f and f is filling, Proposition 5.12 implies that the connected components of $M - g(S)$ are open handlebodies.

However, since g is eventually filling, its Johansson diagram fills S and the boundary of a regular neighbourhood of $g(S)$ is the disjoint union of finitely many 2-spheres. Therefore, the connected components of $M - g(S)$ are open 3-disks and g is a filling immersion. $\square$

Let M be a homotopy sphere and let Σ be a filling Dehn sphere in M. As we have seen, it is possible to find a Dehn sphere Σ' in M homeomorphic to a filling Dehn sphere of $\mathbb{S}^3$. Hence, Σ' is eventually filling in M. Let $f, f' : \mathbb{S}^2 \to M$ be parametrizations of Σ and Σ' respectively. Since M is a homotopy sphere, both immersions f and f' are null-homotopic, so it is possible to deform one into the other with a sequence of Haken moves.

Corollary 5.16 (cf. [31]). *If we can avoid finger moves ± 0 in the deformation of f into f', then M is $\mathbb{S}^3$.* $\square$

5.4   Diagram moves

Let Σ be a Dehn surface in M. Let $f : S \to M$ and $\mathcal{D}$ be a parametrization and the Johansson diagram of Σ, respectively. Until the end of the chapter we will assume that the domain S of Σ is connected.

Modifications carried out on f might alter $\mathcal{D}$. We would like to analyze how f-moves on filling immersions affect their Johansson diagrams. By Lemma 5.6 finger moves 0 are excluded from this analysis.

Assume that the immersion f', with Johansson diagram $\mathcal{D}'$, is obtained from f by a Haken move ϕ which is not a finger move 0. Figures 5.11 to 5.13 show the effect on $\mathcal{D}$ of the different Haken moves. These are the *diagram Haken moves*, and they are called with the same names as their corresponding moves for immersions. For each of these figures, we can assume that there is a 3-disk B in M such that the modification depicted in the right hand side of the figure takes place inside B and $f^{-1}(B)$ is the union of the 2-disks depicted in the left hand side of the same figure. With this assumption, $\mathcal{D}$ and $\mathcal{D}'$ agree outside $f^{-1}(B)$.

We construct a set $\mathcal{N}(\mathcal{D})$ of neighbouring curves of $\mathcal{D}$ like in Chap. 3. Since f is 2-sided and realizes $\mathcal{D}$, there is a thickening $F : S \times [-1, 1] \to M$ of f such that

$$|\mathcal{N}(\mathcal{D})| = f^{-1}(S \times \{-1, 1\}).$$

This thickening F splits $\mathcal{N}(\mathcal{D})$ into two groups: the curves of the bottom sheet, composed by the curves of $\mathcal{N}(\mathcal{D})$ whose image under f lies on the surface $F(S \times \{-1\})$, and the curves of the top sheet, composed by those whose image under f lies on $F(S \times \{1\})$. The modification of f into f' can be extended to a modification of F into a thickening F' of f'. Thus, we can assume that the set $\mathcal{N}(\mathcal{D}')$ of neighbouring curves of $\mathcal{D}'$ agrees with $\mathcal{N}(\mathcal{D})$ outside $f^{-1}(B)$, and that it inherits from $\mathcal{N}(\mathcal{D})$ the splitting into bottom and top sheets.

Saddle move. If ϕ is a saddle move it is equivalent to push Σ along a 2-gon Δ in M as in Figs. 5.9 and 5.11(b) (this operation is explained in detail in Sec. 6.3.3). The "vertices" of Δ (denoted 1 and 2 in Fig. 5.11(b)) are double points of Σ lying on the double curves $\bar{\alpha}$ and $\bar{\beta}$, respectively. If α and $\tau\alpha$ are the lifts of $\bar{\alpha}$ under f, and β and $\tau\beta$ are the lifts of $\bar{\beta}$, then $f^{-1}(\Delta)$ is the union of two arcs a and b in S verifying (Fig. 5.11(a))

$$a \cap |\mathcal{D}| = \partial a, \qquad\qquad b \cap |\mathcal{D}| = \partial b.$$

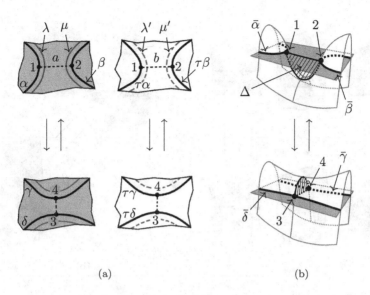

Fig. 5.11: Saddle move.

The arc a (resp. b) meets $|\mathcal{N}(\mathcal{D})|$ at two interior points of a (resp. b), where it crosses the neighbouring curves λ and μ of $\mathcal{D}$ (resp. λ' and μ'). As it can be seen in Fig. 5.11(b), the neighbouring curves λ and μ are in the same sheet. Similarly, λ' is in the same sheet as μ'. After pushing f along Δ, the curves α and β become connected along a, and $\tau\alpha$ and $\tau\beta$ become connected along b. Therefore, $\mathcal{D}'$ is the diagram in the bottom of Fig. 5.11(a).

Finger move 1. Let ϕ be a finger move $+1$ (Fig. 5.12(b)). A new double curve $\bar{\nu}$ and two new triple points 1 and 2 appear in Σ after applying ϕ. The changes introduced by ϕ in $\mathcal{D}$ appear in Fig. 5.12(a) viewed from top to bottom. The diagram $\mathcal{D}'$ has two more curves and six more crossings than $\mathcal{D}'$: the lifts ν and $\tau\nu$ of $\bar{\nu}$ under f' and the triplets of the new triple points 1 and 2. It is trivial to decide how the six new crossings of $\mathcal{D}'$ must be distributed into the triplets of 1 and 2. To decide how the curves ν and $\tau\nu$ should be identified, we must look at the neighbouring curves of $\mathcal{D}'$ that ν and $\tau\nu$ intersect. The curve ν intersects two opposite curves of $\mathcal{N}(\mathcal{D}')$ and one of them, say λ, is in the top sheet. In the same way, let μ be the neighbouring curve of the top sheet intersecting $\tau\nu$. The identification of ν and $\tau\nu$ must be done in such a way that the neighbouring points in $|\nu| \cap |\lambda|$ are related by $\mathcal{D}'$ with the neighbouring points in $|\tau\nu| \cap |\mu|$ (Fig. 5.12(a)).

If ϕ is a finger move -1, we have the converse situation (Fig. 5.12 from

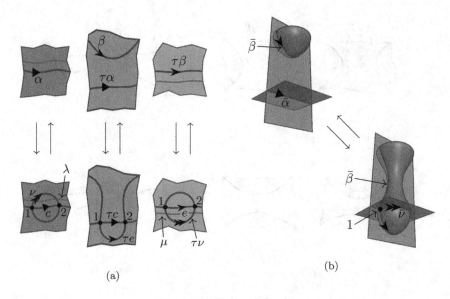

Fig. 5.12: Finger move 1.

bottom to top). There is a pair of disjoint sister curves, ν and $\tau\nu$ of $\mathcal{D}$ which are simple and each of them contains two crossings of $\mathcal{D}$. The curve ν surrounds a closed disk $D_\nu \subset S$ and there is an arc $c \subset |\mathcal{D}|$ such that

$$D_\nu \cap |\mathcal{D}| = |\nu| \cup c.$$

The curve $\tau\nu$ has a similar configuration with disk $D_{\tau\nu}$ and arc e. The union of the sister arcs τc and τe of c and e, respectively, in $\mathcal{D}$ is the boundary of a 2-gon in S. After the finger move -1 on $\mathcal{D}$ the curves ν and $\tau\nu$ and the 2-gon bounded by $\tau c \cup \tau e$ disappear as in Fig. 5.12 (from bottom to top).

Finger move 2. Similar arguments as above apply in this case. The effect on $\mathcal{D}$ of a finger move 2 is depicted in Fig. 5.13.

In general, it is very difficult to visualize the immersion f, and it is even more difficult to visualize the moves performed on it. Since it is easier to work with diagrams, we are interested in the converse construction: we want to control the Haken moves on f looking only at the diagram $\mathcal{D}$. Assume that $\mathcal{D}$ is modified by a Haken move $\phi_\mathcal{D}$ as in one of the Figs. 5.11(a), 5.12(a) or 5.13(a), obtaining a new diagram $\mathcal{D}'$.

Definition 5.17. The diagram Haken move $\phi_\mathcal{D}$ is *realizable from* f if f can be modified by a Haken move ϕ whose result in $\mathcal{D}$ is $\phi_\mathcal{D}$.

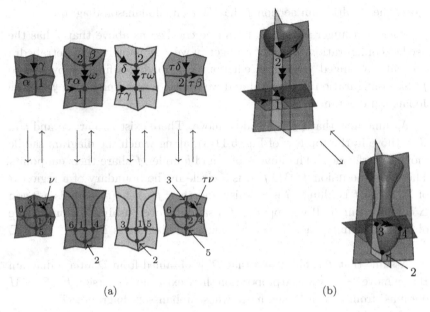

Fig. 5.13: Finger move 2.

Theorem 5.18. *If f is a filling immersion, any Haken move on $\mathcal{D}$ is realizable from f.*

Proof. Assume that f is filling. If $\phi_{\mathcal{D}}$ is a finger move $+1$ or $+2$, it is obvious that we can apply to f a finger move of the same type to obtain a immersion whose Johansson diagram is $\mathcal{D}'$.

If $\phi_{\mathcal{D}}$ is a finger move -1, then we started with a configuration similar to the one in the bottom row of Fig. 5.12(a) and we moved to a situation as the one in the top row of the same figure. The small disk $D_\nu \subset S$ bounded by $|\nu|$ is divided by c into two 2-gons Δ_1 and Δ_2. Assume that Δ_1 is the 2-gon whose boundary is formed by the two arcs indicated with arrows in Fig. 5.12(a). On the other hand, the disk $D_{\tau\nu} \subset S$ bounded by $\tau\nu$ is divided by e into two 2-gons Δ_3 and Δ_4. Suppose also that Δ_3 is the 2-gon bordered by the two arcs indicated by arrows in Fig. 5.12(a). Let $\Delta_5 \subset S$ be the 2-gon bounded by $\tau c \cup \tau e$. For each $i = 1, \ldots, 5$, define $\overline{\Delta}_i = f(\Delta_i)$. Since f is filling, the 2-gons $\overline{\Delta}_1$, $\overline{\Delta}_3$ and $\overline{\Delta}_5$ are the three faces of a region of Σ which is a trihedron, and the same happens with $\overline{\Delta}_2$, $\overline{\Delta}_4$ and $\overline{\Delta}_5$. Hence, there exist two trihedral regions of Σ with a common face, so it is possible to apply to f a finger move -1 to make both trihedra disappear. After such move, the resulting immersion f' has $\mathcal{D}'$ as its Johansson diagram.

If $\phi_{\mathcal{D}}$ is a finger move -2, it can be checked as above that f has the needed configuration to apply a finger move -2: namely, four tetrahedra cyclically arranged around an edge of Σ. Applying a finger move -2 to f this configuration is removed and we obtain a new immersion f' whose Johansson diagram is $\mathcal{D}'$.

Assume now that $\phi_{\mathcal{D}}$ is a saddle move. There exist two arcs a and b in S as those in the top row of Fig. 5.11(a), along which the diagram saddle move is performed. The images of a and b under f share their endpoints. Therefore the union $f(a) \cup f(b)$ is a circle in the boundary of a region R of Σ. Since f is filling, R is a 3-disk and then $f(a) \cup f(b)$ bounds a 2-gon Δ contained in R. If we apply to f a saddle move pushing Σ along Δ, we obtain a new immersion f' whose Johansson diagram is $\mathcal{D}'$. $\qquad\square$

Assume that f is filling and that $\mathcal{D}'$ is obtained from $\mathcal{D}$ after a diagram Haken move $\phi_{\mathcal{D}}$. By above proposition there exists an immersion $f' : S \to M$ obtained from f by a Haken move whose Johansson diagram is $\mathcal{D}'$.

Proposition 5.19. *If $\mathcal{D}'$ is filling, f' is filling.*

Proof. Call $\Sigma' = f'(S)$. Since $\mathcal{D}'$ is filling, it suffices to prove that each connected component of $M - \Sigma'$ is an open 3-disk. Finger moves ± 1 and ± 2 do not modify the topological type of the regions of Σ. By Lemma 3.20, there is a regular neighbourhood of Σ' whose boundary is a disjoint union of 2-spheres. By Proposition 5.12, if $\phi_{\mathcal{D}}$ is a saddle move, all the regions of Σ' are open handlebodies. Therefore, all the regions of Σ' are open 3-disks. $\qquad\square$

Theorem 5.20. *If the diagrams $\mathcal{D}$ and $\mathcal{D}'$ on S are filling and they are related by a Haken move, then $M(\mathcal{D}) = M(\mathcal{D}')$.*

Proof. Let $f : S \to M(\mathcal{D})$ be a filling immersion realizing $\mathcal{D}$. By Theorem 5.18, there exists an immersion $f' : S \to M(\mathcal{D})$ obtained from f by a Haken move and whose Johansson diagram is $\mathcal{D}'$. By Proposition 5.19, f' is filling, and by Theorem 3.23 we conclude that $M(\mathcal{D}) = M(\mathcal{D}')$. $\qquad\square$

In this way, most of the results given for filling immersions have a natural translation to Johansson diagrams. If $\mathcal{D}$ is a filling diagram and it is modified by a Haken move, this move will be an f-move if the resulting diagram is filling. By Lemma 5.6, finger moves $+1$ and ± 2 on diagrams are always

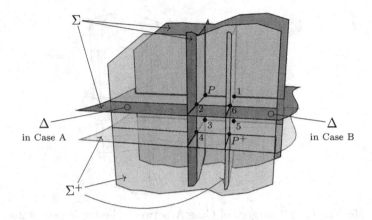

Fig. 5.14: Local configuration of Σ and Σ^+ near P.

f-moves, while finger moves -1 and saddle moves might fail to be filling preserving.

Definition 5.21. Two filling diagrams on a surface S are *f-equivalents* if it is possible to transform one into the other using a finite sequence of f-moves and ambient isotopies of S.

A filling diagram $\mathcal{D}$ is *null-homotopic* if there exists a null-homotopic filling immersion $f : S \to M(\mathcal{D})$ realizing $\mathcal{D}$. In fact, if $\mathcal{D}$ is null-homotopic any filling immersion $g : S \to M(\mathcal{D})$ realizing $\mathcal{D}$ is null-homotopic. Next result follows from Theorems 5.8 and 5.20.

Theorem 5.22. *Let $\mathcal{D}$ and $\mathcal{D}'$ be two null-homotopic filling diagrams on $\mathbb{S}^2$. Then $M(\mathcal{D}) = M(\mathcal{D}')$ if and only if $\mathcal{D}$ and $\mathcal{D}'$ are f-equivalents.* □

Remark 5.23. This theorem gives a complete system of moves for null-homotopic Johansson representations on $\mathbb{S}^2$ of 3-manifolds.

5.5 Duplication

It would be interesting to have an algorithm to determine whether a filling diagram on $\mathbb{S}^2$ is null-homotopic or not. Currently we ignore the existence of such algorithm. However, [69] introduces the *duplication algorithm* that, given a filling diagram $\mathcal{D}$ on $\mathbb{S}^2$, provides a null-homotopic filling diagram $\mathrm{dup}(\mathcal{D})$ on $\mathbb{S}^2$ representing the same 3-manifold as $\mathcal{D}$. The diagram $\mathrm{dup}(\mathcal{D})$ is called the *duplicate* of $\mathcal{D}$.

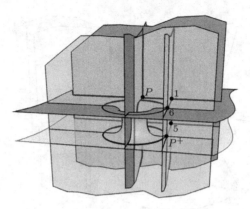

Fig. 5.15: The type 2 surgery of Case A between the triple points 2 and 4.

Following [69], given a filling Johansson diagram $\mathcal{D}$ on the 2-sphere the steps of the algorithm are:

(Step 1) Construct the 3-manifold $M(\mathcal{D})$ and the Dehn surface $\Sigma \subset M(\mathcal{D})$.

(Step 2) Since $\mathcal{D}$ is in $\mathbb{S}^2$ there exists at least 4 distinct regions of Σ, and therefore, there is a face $f(\Delta) \subset \Sigma$ incident with different regions.

(Step 3) Choose a triple point $P \in \mathrm{cl}(\Delta) \subset \mathbb{S}^2$.

(Step 4) Take a thickening F of a parametrization f of Σ. Define

$$\Sigma^+ = F(\mathbb{S}^2 \times \{1\}).$$

The union $\Sigma \cup \Sigma^+$ is a filling pair of spheres, whose diagram is formed by two copies of $\mathcal{D}$ (one in each sphere) with parallel curves coming from $\Sigma \cap \Sigma^+$ (see Fig. 5.14). Replacing F with $F'(\cdot, t) = F(\cdot, -t)$ if necessary, one of the following two possibilities occur:

(A) from the two parallel curves to the curves forming $\partial \Delta$ near P, one is outside Δ parallel to an edge $\bar{e}$, and the other one runs inside Δ.

(B) the two parallel curves go inside Δ.

Figure 5.16 shows the two possible configurations around the triplet of the triple point P.

Now, depending on the case:

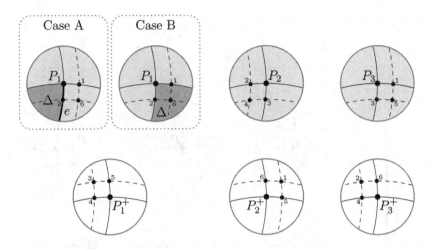

Fig. 5.16: Local configuration around the triplets of $P \in \Sigma$ (top row) and $P^+ \in \Sigma^+$ (bottom row). Top disks are parallel to the bottom ones. Dashed lines correspond to double curves inside $\Sigma \cap \Sigma^+$ while solid ones are double curves of the diagram $\mathcal{D}$ in Σ and Σ^+. Compare with Fig. 5.14.

(Step 5A) Consider the triplet $\{P_1, P_2, P_3\}$ of P in the original sphere and assume $P_1 \in \mathrm{cl}(f^{-1}(\Delta))$. There is another double point in the edge e of $\mathcal{D}$ incident with P_1 and $f(e) = \bar{e}$ which corresponds to the triple point in $2 \in \Sigma \cap \Sigma^+$. The situation is like in Fig. 5.16.

(Step 6A) Make a simple Type 2 surgery (Sec. 2.4.3) between the double points 2 and 4, see Figs. 5.14, 5.15 and 5.17: remove little disks around 2 and 4 and glue the boundary of the disks (the gray curves in the picture) starting from the starred point following the direction of the arrow.

(Step 5B) In this case, there is a double point in the interior of Δ near P_1. This double point belongs to the triplet of the triple point $6 \in \Sigma \cap \Sigma^+$.

(Step 6B) Finally, do a simple Type 2 surgery between P^+ and 6, as in Step 6-a following the instructions of Case B in Fig. 5.17.

Theorem 5.24. *The filling Dehn sphere obtained from* $\mathrm{dup}(\mathcal{D})$ *is filling and null-homotopic.*

Proof. To prove that $\mathrm{dup}(\mathcal{D})$ is filling it is enough to see that the piping in Step 6 does not connect a region with itself. In Case B, it is trivial, since Δ

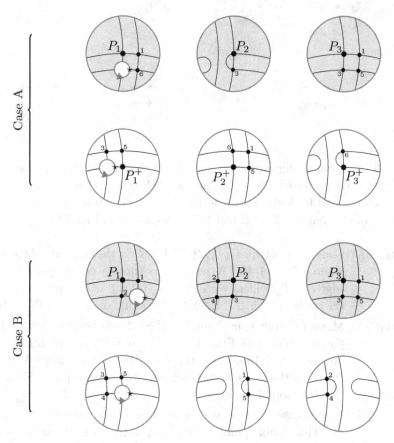

Fig. 5.17: Local configuration around the triplets of $P \in \Sigma$ (gray disks) and $P^+ \in \Sigma^+$ (white disks) after the removal of a little disk and before the simple type 2 surgery.

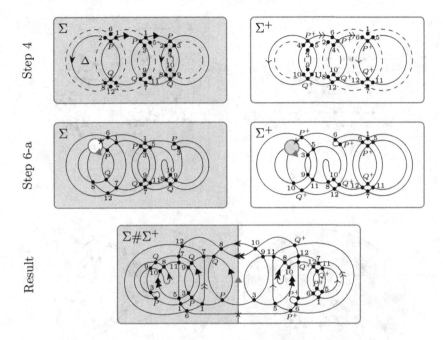

Fig. 5.18: Duplication applied to a filling Dehn sphere in $\mathbb{S}^2 \times \mathbb{S}^1$.

separates different regions of Σ. In Case A, half-the-tube makes connections between the two regions that Δ separates. The other half makes connections between regions laying between Σ and Σ^+, but these regions run alongside the faces. The faces meeting the edge $\bar{e}$ are distinct, because Δ lays between two distinct regions. Therefore, the parallel regions to these faces must be distinct.

On the other hand, in either case, the filling Dehn sphere $\Sigma \# \Sigma^+$ has as a parametrization $f_\#$ the thickening F restricted to a sphere built with $\mathbb{S}^2 \times \{1\}$ and $\mathbb{S}^2 \times \{0\}$ with two disks removed and a tube gluing the resulting boundaries. Hence, $f_\#$ is the restriction of an immersion of a 3-ball and therefore null-homotopic. □

Example 5.25. Consider the filling Dehn sphere in $\mathbb{S}^2 \times \mathbb{S}^1$ of Sec. 7.2.1 whose Johansson diagram is depicted in the left of the top row of Fig. 5.18. It is easy to check that the face labelled as Δ in the top left diagram of Fig. 5.18 separates two regions of the corresponding Dehn surface. Then, Case A should be applied to compute its duplicate. The construction of the duplicate is depicted in Fig. 5.18.

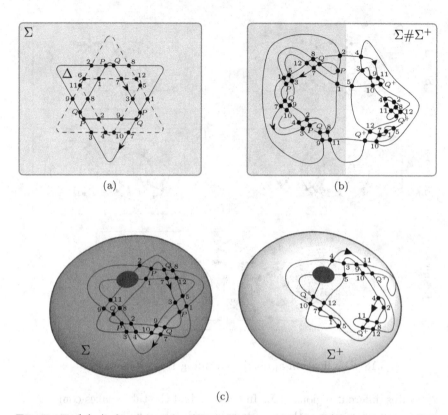

(c)

Fig. 5.19: (a) A diagram of a filling Dehn sphere in $L(3,1)$ with parallel curves (dashed), (b) the duplicate of the former diagram, and (c) the two domains before the connected sum.

Consider now the example of Sec. 7.2.3. It is possible to check that Δ in Fig. 5.19(a) separates two regions of the corresponding Dehn surface in $L(3,1)$. (Since $L(3,1)$ is irreducible, this example is null-homotopic and duplication is unnecessary. We use it just to illustrate the algorithm). Therefore, we are in Case B of the duplication algorithm. The piping done on the Dehn surfaces corresponds to gluing the two spheres depicted in Fig. 5.19(c). Figure 5.19(b) shows the duplicate.

As a corollary of previous theorem we have the following result:

Theorem 5.26. *Two filling diagrams $\mathcal{D}$ and $\mathcal{D}'$ on $\mathbb{S}^2$ represent the same 3-manifold if and only $\mathrm{dup}(\mathcal{D})$ and $\mathrm{dup}(\mathcal{D}')$ f-equivalents.* $\square$

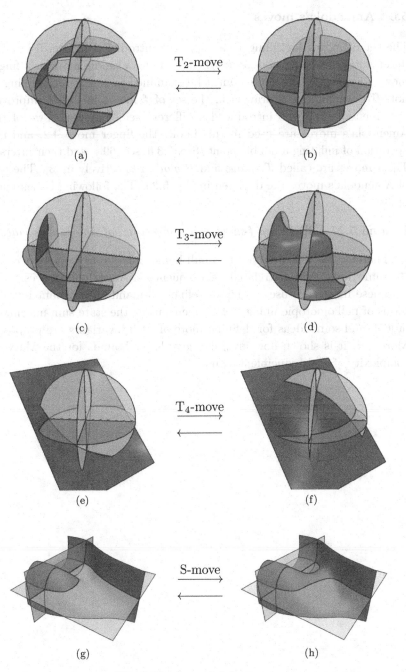

(a)   T$_2$-move   (b)

(c)   T$_3$-move   (d)

(e)   T$_4$-move   (f)

(g)   S-move   (h)

Fig. 5.20: Amendola's moves.

5.6   Amendola's moves

The set of moves for filling Dehn surfaces introduced in Sec. 5 has the disadvantage of being *non-local*: when applying a saddle move or a finger move -1 the resulting surface must be examined to check if it is filling or not. To overcome this drawback, the set of f-moves has been improved by Amendola [3], who introduced a different set of moves. Two of the Amendola's moves are used in this book: the finger move $+2$, and the operation of inflating a double point (Sec. 6.3.5, see [69]), and their inverses. These moves are called T_1-*move* and *B-move*, respectively in [3]. The rest of Amendola's moves are depicted in Fig. 5.20. The following is Lemma 6 of [3]).

Lemma 5.27 ([3]). *Each f-move is a composition of Amendola's moves.*

Therefore by Theorem 5.8 two null-homotopic filling Dehn spheres of the same 3-manifold are related by a sequence of Amendola's moves.

These moves are used in [3] to define a 3-manifold invariant inv_m in terms of nulhomotopic filling Dehn spheres using the state sum machinery of [65], and some ideas for defining more of such invariants are proposed. Moreover, it is shown how inv_m can give lower bounds for the Matveev complexity of $\mathbb{P}^2$-irreducible 3-manifolds.

Chapter 6

Proof of Theorem 5.8

6.1 Pushing disks

In this section all surfaces and 3-manifolds are assumed to be closed and orientable unless otherwise stated. Let S be a surface and M a 3-manifold.

Let $f, g : S \to M$ be two immersions such that there exists a closed disk $D \subset S$ verifying:

(1) f and g agree on $S - D$;

(2) $f|_D$ and $g|_D$ are embeddings;

(3) $f(D)$ and $g(D)$ meet exactly at $f(\partial D) = g(\partial D)$; and

(4) $f(D) \cup g(D)$ is the boundary of a 3-disk B in M (Fig. 6.1).

The pair (D, B) is a *pushing disk* (see [32]) between f and g, or, equivalently, it is said that g *is obtained from f by pushing the disk D along B*. In the pushing disk (D, B), D is the *pushed disk* and the B is the *pushing ball*. The common image $f(\partial D) = g(\partial D)$, denoted by eq($B$), is the *equator* of B. If f and g are transverse immersions (D, B) is *transverse*. Given a pushing disk (D, B), there is no restriction on how the rest of the immersed surface $f(S - D) = g(S - D)$ intersects B (Fig. 6.1(b)).

Remark 6.1. Homma-Nagase and Haken moves are particular cases of pushing disks.

An interesting tool in this context is the following lemma due to Smale.

Lemma 6.2 ([64]). *Any diffeomorphism of the closed disk fixing the boundary is isotopic to the identity by a boundary-fixing isotopy.* □

A consequence of this lemma is that, given a map f and a pushing disk (D, B), the map g is completely determined up to ambient isotopy in S.

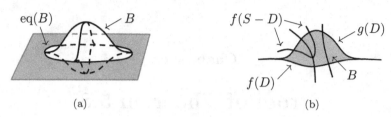

(a)   (b)

Fig. 6.1: A pushing disk.

Remark 6.3. If $g, g' : S \to M$ are obtained from f and the same pushing disk (D, B), then g and g' are ambient isotopic in S by an ambient isotopy supported by D.

Two (transverse) immersions $f, g : S \to M$ are *regularly homotopic by (transverse) pushing disks* if there exists a finite sequence

$$f = f_0, f_1, \ldots, f_n = g$$

of (transverse) immersions such that f_i is obtained from f_{i-1} by a pushing disk, for any $i = 1, \ldots, n$. Similarly, two embeddings f and g are *isotopic by pushing disks* if there exists a finite sequence of embeddings $f = f_0, f_1, \ldots, f_n = g$ such that f_i is obtained from f_{i-1} by a pushing disk, for any $i = 1, \ldots, n$.

Definition 6.4. A regular homotopy $H : S \times I \to M$ such that H_0 and H_1 are transverse is a regular homotopy *with isolated singularities* if there exists $t_1, t_2, \ldots, t_n \in I$ with

$$0 < t_1 < t_2 < \cdots < t_n < 1$$

such that

(i) H_t is transverse for all $t \in I - \{t_1, \ldots, t_n\}$; and
(ii) for each $i = 1, 2, \ldots, n$, the points of M where H_{t_i} is not transverse form a discrete set.

By Lemma 1.2, H as in the definition above is a n ambient isotopy when restricted to any subinterval of $I - \{t_1, \ldots, t_n\}$. It is clear that if a regular homotopy $H : S \times I \to M$ is decomposed into Haken and/or Homma-Nagase moves and ambient isotopies, it is a regular homotopy with isolated singularities. Then, by Theorem 5.3:

Corollary 6.5. *If two transverse immersions $f, g : S \to M$ are regularly homotopic, there exists a regular homotopy with isolated singularities connecting f and g.* $\qquad\square$

The first step of the proof of Theorem 5.8 is the following result:

Theorem 6.6. *Let $f, g : S \to M$ be two regularly homotopic immersions. Then:*

(A) they are regularly homotopic by pushing disks;

(B) if both are transverse, they are regularly homotopic by transverse pushing disks; and

(C) if f and g are regularly homotopic by a regular homotopy with isolated singularities fixing a compact subsurface $S' \subset S$, then f and g are regularly homotopic by transverse pushing disks fixing S'.

A *compact subsurface* of the surface S is a compact, perhaps non-connected, surface with boundary embedded in S.

In this way, a regular homotopy can be decomposed into a finite sequence of pushing disks. While Homma-Nagase and Haken moves are particular cases of pushing disks, theorem above is not a weaker version of Theorem 5.3. Theorem 5.3 implies that any regular homotopy between transverse immersions can be decomposed into transverse pushing disks *and ambient isotopies*. Decomposing these ambient isotopies into pushing disks keeping the requirements of item C in the statement is the hardest part of the proof of Theorem 6.6.

The remainder of this section is devoted to the proof of Theorem 6.6. This proof is mainly inspired by [35]. In particular, the next result comes from Lemma 11 of [35].

Lemma 6.7. *Let $f, g : S \to M$ be two immersions. Assume that there exists a closed disk $D \subset S$ such that:*

(i) f and g agree on $S - D$;

(ii) $f|_D$ and $g|_D$ are embeddings; and

(iii) $f(D)$ and $g(D)$ are contained in the interior of a 3-ball B_* in M.

Then, f is obtained from g by two pushing disks with pushed disk D. If both f and g are transverse immersions, f is obtained from g by two transverse pushing disks with pushed disk D.

Proof. Since $f|_D$ and $g|_D$ are embeddings, $\overline{D}_0 = f(D)$ and $\overline{D}_1 = g(D)$ are embedded disks in the interior of B_*. By continuity, the immersions f and g also agree on $\mathrm{cl}(S - D)$, and since S has no boundary, $\partial \overline{D}_0 = \partial \overline{D}_1$.

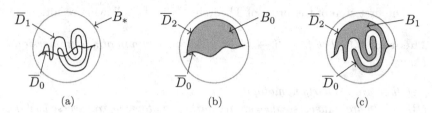

$$(a) \qquad\qquad (b) \qquad\qquad (c)$$

Fig. 6.2: The idea of the proof of Lemma 6.7.

There is a third disk $\overline{D}_2$ in the interior of B_* such that:

(i) $\overline{D}_2 \cap \overline{D}_0 = \overline{D}_2 \cap \overline{D}_1 = \partial \overline{D}_0 = \partial \overline{D}_1 = \partial \overline{D}_2$; and

(ii) $\overline{D}_2 = h(D)$,

where $h : S \to M$ is a third immersion that also agrees with f and g on $S - D$.

The union $\overline{D}_2 \cup \overline{D}_0$ is homeomorphic to a 2-sphere contained in the interior of B_*, so it is the boundary of a 3-disk B_0 in the interior of B_*. Similarly, the union $\overline{D}_2 \cup \overline{D}_1$ encloses a closed 3-disk B_1 contained in the interior of B_*. This means that h is obtained from f by the pushing disk (D, B_0), and that g is obtained from h by the pushing disk (D, B_1). Therefore, f is obtained from g by means of two pushing disks (Fig. 6.2). If f and g are transverse D_2 can be chosen such that h is also transverse. □

When applying this Lemma, the map h introduced in the proof is called *intermediate map between f and g*.

If $H : S \times I \to M$ is a regular homotopy, its track $\widehat{H}$ is an immersion. Similarly, if H is an isotopy, $\widehat{H}$ is an embedding. Likewise an immersion looks locally like an embedding, a regular homotopy looks locally like an isotopy. In particular, we have the following lemma.

Lemma 6.8. *Let $H : S \times I \to M$ be a regular homotopy. For each $x \in S$, there exists a neighbourhood $U_x \subset S$ of x in S such that the restriction $H|_{U_x \times I}$ is an isotopy.*

Proof. Consider the distance d_S induced by a Riemannian metric on S. Take the distance $d_{S \times I}$ on $S \times I$ defined as $d_{S \times I}(s, t) = d_S(s) + d_i(t)$ where d_i is the standard distance on I.

Since $\widehat{H}$ is an immersion, there is an open cover $\widehat{\mathcal{U}}$ of $S \times I$ such that $\widehat{H}|_{\widehat{U}}$ is an embedding for each $\widehat{U} \in \widehat{\mathcal{U}}$.

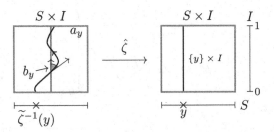

Fig. 6.3

By compactness of S, the cover $\widehat{\mathcal{U}}$ has Lebesgue number $\delta > 0$ for the distance $d_{S \times I}$. Now, for each $x \in S$ take U_x as the ball (in the distance d_S) centered at x with radius δ. For all $t \in I$, the product $U_x \times \{t\}$ is contained in an element of $\widehat{\mathcal{U}}$, so $\widehat{H}|_{U_x \times \{t\}}$ is an embedding. This implies that $H_t|_{U_x}$ is an embedding for all $t \in I$ and, therefore, $H|_{U_x \times I}$ is an isotopy. $\qquad \square$

Let $H : S \times I \to M$ be a regular homotopy. A (open or closed) disk or subsurface $U \subset S$ is *isotopically deformed by* H if the restriction $H|_{U \times I}$ is an isotopy. A cover $\mathcal{U}$ of S is *isotopically deformed by* H if so are its elements.

Lemma 6.9. *Let S be a surface, let $\zeta : S \times I \to S$ be an ambient isotopy of S and let $\mathcal{U}$ be an open cover of S. Then, there exists a collection $\zeta^1, \ldots, \zeta^n$ of ambient isotopies of S such that:*

(1) for each $t \in I$, $\zeta_t = \zeta_t^n \circ \cdots \circ \zeta_t^1$; and
(2) the support of each ζ^i is contained in an element of $\mathcal{U}$.

Moreover, if ζ fixes a set $Y \subset S$, the ambient isotopies $\zeta^1, \ldots, \zeta^n$ also fix Y.

A purely topological version of this result can be found in [15, Corollary 1.3]. The proof shown here is a translation to the smooth case of the proofs of Theorem I and Addendum I.I of [35].

Proof. Consider the track $\widehat{\zeta} : S \times I \to S \times I$ of ζ given by

$$\widehat{\zeta}(x,t) = (\zeta(x,t),t).$$

Since ζ is an ambient isotopy of S, the map $\widehat{\zeta}$ is a level-preserving diffeomorphism of $S \times I$.

Following [35] we need to measure angles in $S \times I$. To do so, fix an arbitrary Riemannian metric on S, and consider the product metric on

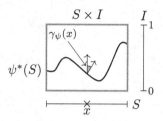

Fig. 6.4: The map γ_ψ.

$S \times I$ when I is endowed with its standard metric. Tangent vectors of $S \times I$ parallel to the S factor are *horizontal*, and vectors parallel to the I factor are *vertical*.

For $y \in S$, consider the arc $a_y = \widehat{\zeta}^{-1}(\{y\} \times I)$ in $S \times I$. Then

$$a_y = \left\{ (\zeta_t^{-1}(y), t) \mid t \in I \right\}.$$

Let β_y be the maximal angle between a_y and the vertical component of $S \times I$, considering both pointing "upwards", that is, orienting I from 0 to 1, as in Fig. 6.3 [1]. Since $\widehat{\zeta}$ is level-preserving, the tangent vectors to a_y in $S \times I$ are never horizontal, and therefore $\beta_y < \frac{\pi}{2}$ for all $y \in S$. By construction, β_y depends continuously on y, and since S is compact, it has a maximal value $\beta < \frac{\pi}{2}$.

Given a map $\psi : S \to I$, consider its graph $\psi^* : S \to S \times I$ given by $\psi^*(x) = (x, \psi(x))$. For each $x \in S$, let $\gamma_\psi(x)$ be the angle between the normal vector to $\psi^*(S)$ at $\psi^*(x)$ and the vertical component of $S \times I$, both "pointing upwards" (Fig. 6.4). Since ψ^* is a graph map and ψ is smooth, $\gamma_\psi(x) < \frac{\pi}{2}$ for any $x \in S$. By compactness, there is an $x \in S$ where the map γ_ψ reach its maximum $\bar{\gamma}_\psi$, which obviously satisfy $\bar{\gamma}_\psi < \frac{\pi}{2}$. Given an angle $\delta \in [0, \frac{\pi}{2})$, the map ψ is δ-*flat* if $\bar{\gamma}_\psi \leq \delta$. Let denote by $\mathcal{F}_\delta$ the set of δ-flat maps from S to I. Let us make some remarks on δ-flat maps:

(i) $\mathcal{F}_0$ only contains constant maps.
(ii) Given a map $\psi : S \to I$ and $t \in I$, the function $t\psi$ has a "flatter" graph than ψ, and therefore $\bar{\gamma}_{t\psi} \leq \bar{\gamma}_\psi$. In particular, if ψ is δ-flat for some $\delta \in [0, \frac{\pi}{2})$ so is $t\psi$ for any $t \in I$. Moreover, for any $\delta' \in [0, \frac{\pi}{2})$, the map $t\psi$ is δ'-flat for a sufficiently small t.

---

[1]We orient the arc a_y and all the arcs $\{x\} \times I$ as I. For each $t \in I$, we define $\beta_y(t)$ as the angle between the tangent vector of a_y and another vector tangent to $\{\zeta_t^{-1}(y)\} \times I$, assuming that both vectors are positive with respect to the fixed orientations. The value β_y is the maximal value of the function $\beta_y(t)$ in I.

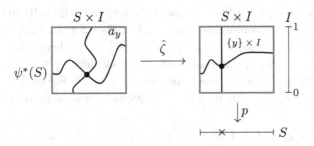

Fig. 6.5

(iii) Fix $\delta \in [0, \frac{\pi}{2} - \beta)$ and consider $\psi \in \mathcal{F}_\delta$. Since $\delta + \beta < \frac{\pi}{2}$, the graph $\psi^*(S)$ of ψ transversely meets each arc a_y in a unique point (see [35, p. 77] and Fig. 6.5). This implies that the image of $\psi^*(S)$ under $\widehat{\zeta}$ has a unique intersection point with the sets of the form $\{y\} \times I$, and this intersection is transverse. If $p : S \times I \to S$ is the first coordinate projection, all this implies that p is a diffeomorphism from the submanifold $(\widehat{\zeta} \circ \psi^*)(S)$ onto S, and hence $p \circ \widehat{\zeta} \circ \psi^* : S \to S$ is a diffeomorphism.

Claim 6.10. *There exists a finite sequence of maps $\psi_0, \psi_1, \ldots, \psi_n \in \mathcal{F}_\delta$ such that $\psi_0 \equiv 0$, $\psi_n \equiv 1$ and ψ_{i-1} and ψ_i agree on S except in an element of $\mathcal{U}$, for all $i = 1, \ldots, n$.*

Proof of Claim. Let Θ be a smooth partition of unity subordinate to $\mathcal{U}$ [30, p. 43]. Since S is compact, Θ can be assumed to be finite.

Then $\Theta = \{\theta_1, \ldots, \theta_k : S \to I\}$ is a collection of maps such that:

(1) for each $j = 1, \ldots, k$, there exists $U \in \mathcal{U}$ such that
$$\mathrm{supp}(\theta_j) = \mathrm{cl}(\{x \in S \mid \theta_j(x) \neq 0\}) \subset U;$$
and
(2) $(\theta_1 + \cdots + \theta_k)(x) = 1$ for all $x \in S$.

There is a sufficiently large $q \in \mathbb{N}$ such that $\frac{1}{q}(\theta_1 + \cdots + \theta_j)$ is δ-flat for all $j = 1, \ldots, k$. In this case, the collection of maps

$$0, \tfrac{1}{q}\theta_1, \tfrac{1}{q}(\theta_1 + \theta_2), \ldots, \tfrac{1}{q}(\theta_1 + \cdots + \theta_{k-1}), \tfrac{1}{q},$$

$$\tfrac{1}{q} + \tfrac{1}{q}\theta_1, \tfrac{1}{q} + \tfrac{1}{q}(\theta_1 + \theta_2), \ldots, \tfrac{1}{q} + \tfrac{1}{q}(\theta_1 + \cdots + \theta_{k-1}), \tfrac{2}{q},$$

$$\cdots$$

$$\tfrac{q-1}{q}, \tfrac{q-1}{q} + \tfrac{1}{q}\theta_1, \tfrac{q-1}{q} + \tfrac{1}{q}(\theta_1 + \theta_2), \ldots, \tfrac{q-1}{q} + \tfrac{1}{q}(\theta_1 + \cdots + \theta_{k-1}), 1$$

satisfy the statement. $\square$

Let $\psi_0, \ldots, \psi_n \in \mathcal{F}_\delta$ be as in the statement of Claim 6.10, and consider the collection of diffeomorphisms $k_0, \ldots, k_n : S \to S$ given by $k_i = p \circ \widehat{\zeta} \circ \psi_i^*$, for $i = 0, \ldots, n$. Note that $\zeta_0 = k_0 = \mathrm{id}_S$ and $\zeta_1 = k_n$. For each $i = 1, \ldots, n$, the maps ψ_{i-1} and ψ_i differ, at most, on an element $U_i \in \mathcal{U}$. Then, k_{i-1} y k_i also differ (at most) on the same element $U_i \in \mathcal{U}$. For each $i = 1, \ldots, n$, define $h_i = k_i \circ k_{i-1}^{-1}$. The map h_i is a diffeomorphism supported by U_i, and

$$\zeta_1 = k_n = h_n \circ h_{n-1} \circ \cdots \circ h_1.$$

Now, for each $t \in I$ and $i = 0, \ldots, n$, consider the map $\psi_{i,t} = t\psi_i$. Since $\psi_0, \ldots, \psi_n$ are all δ-flat, $\psi_{0,t}, \ldots, \psi_{n,t}$ will also be δ-flat. Repeating the construction done for $\psi_0, \ldots, \psi_n$, we obtain $k_{i,t} = p \circ \widehat{\zeta} \circ \psi_{i,t}^*$ and $h_{i,t} = k_{i,t} \circ k_{i-1,t}^{-1}$, for each $i = 1, \ldots, n$ and all $t \in I$. For any $t \in I$ and all $i = 1, \ldots, n$, the diffeomorphism $h_{i,t}$ is supported by U_i. Moreover, $k_{0,t} = \mathrm{id}_S$ and $k_{n,t} = \zeta_t$, so

$$\zeta_t = h_{n,t} \circ h_{n-1,t} \circ \cdots \circ h_{1,t}.$$

Let define the ambient isotopies $\zeta^1, \ldots, \zeta^n$ in S by $\zeta^i(x, t) = h_{i,t}(x)$ with $i = 1, \ldots, n$. Then, ζ^i is supported by an element of $\mathcal{U}$ and $\zeta_t = \zeta_t^n \circ \cdots \circ \zeta_t^1$ for all $t \in I$. If the ambient isotopy ζ leaves a subset $Y \subset S$ fixed, all the diffeomorphisms $k_{i,t}$ will also fix the subset Y, and therefore the ambient isotopies $\zeta^1, \ldots, \zeta^n$ also fix Y. □

Being transverse is stable under "small deformations". This idea is formalized in the following result that follows from [20, Chap. III].

Lemma 6.11. *Let $f : N \to N'$ a map between the manifolds N and N', and let $H : N \times I \to N'$ be a homotopy with $H_0 = f$. If f is a transverse immersion, an embedding or a diffeomorphism, there exists $\varepsilon > 0$ such that for all $t \in [0, \varepsilon]$ the map H_t is a transverse immersion, an embedding or a diffeomorphism respectively.* □

An immediate corollary of Lemmas 6.7 and 6.9 is the following result.

Proposition 6.12. *Let $f : S \to M$ be an immersion and let ζ be an ambient isotopy of S. Define $g = f \circ \zeta_1$. Then*

(i) f and g are regularly homotopic by pushing disks;

(ii) if f is transverse, f and g are regularly homotopic by transverse pushing disks; and

(iii) if f is an embedding, f and g are isotopic by pushing disks.

In all these cases, if ζ fixes a compact subsurface $S' \subset S$, the sequence of pushing disks connecting f and g can be chosen fixing S'.

Proof. The map $H = f \circ \zeta$ is a regular homotopy between f and g.

Consider a cover $\mathcal{U}$ of S by open disks isotopically deformed by H and such that $\mathrm{cl}(U)$ is a closed disk for all $U \in \mathcal{U}$. By Lemma 6.9, there exist a family of ambient isotopies $\zeta^1, \ldots, \zeta^n$ of S such that,

$$\zeta_t = \zeta_t^n \circ \cdots \circ \zeta_t^1$$

for all $t \in I$, and each ζ^i is supported by an element $U_i \in \mathcal{U}$. In particular,

$$\zeta_1 = \zeta_1^n \circ \cdots \circ \zeta_1^1.$$

Define $h_0 = \mathrm{id}_S$, $h_i = \zeta_1^i \circ \cdots \circ \zeta_1^1$ for $i = 1, \ldots, n$ and $g_i = f \circ h_i$ for $i = 0, \ldots, n$. For a given $i \in \{1, \ldots, n\}$, the diffeomorphism ζ_1^i of S fixes $S - U_i$. Since $h_i = \zeta_1^i \circ h_{i-1}$, the maps h_i and h_{i-1} agree on all points of $x \in S$ such that $h_{i-1}(x) \notin U_i$. It follows that the immersions g_{i-1} and g_i only differ in the disk $h_{i-1}^{-1}(U_i)$.

On the other hand,

$$g_i(h_{i-1}^{-1}(U_i)) = (f \circ h_i)(h_{i-1}^{-1}(U_i)) = (f \circ \zeta_1^i \circ h_{i-1})(h_{i-1}^{-1}(U_i))$$
$$= (f \circ \zeta_1^i)(U_i) = f(U_i) = (f \circ h_{i-1})(h_{i-1}^{-1}(U_i)) = g_{i-1}(h_{i-1}^{-1}(U_i)),$$

and the disk $f(U_i)$ is embedded in M because $\mathcal{U}$ is isotopically deformed by H. By Lemma 6.7, the map g_i can be obtained from g_{i-1} by two pushing disks. Since i is arbitrary, $f = g_0$ and $g = g_n$ are regularly homotopic by pushing disks.

If f is transverse (embedding), then all the maps g_i are transverse (embeddings) and the intermediate maps between g_{i-1} and g_i can be chosen to be transverse (embeddings) for all $i = 1, \ldots, n$.

If H fixes a compact subsurface $S' \subset M$, it is enough to take the cover $\mathcal{U}$ such that for all $U \in \mathcal{U}$ one the following conditions hold: (i) $U \subset S'$; (ii) $U \subset S - S'$; or (iii) $\mathrm{cl}(U - S')$ is a closed disk. By Lemma 6.9 it can be assumed that S' is fixed by $\zeta^1, \ldots, \zeta^n$.

Repeating the above construction, $h_1, \ldots, h_n$ fix S' and that $g_1, \ldots, g_n$ agree with f on S'. As before, the immersions g_{i-1} and g_i only differ in the disk $h_{i-1}^{-1}(U_i)$, for any $i = 1, \ldots, n$. If $U_i \subset S'$, then U_i is fixed by h_{i-1} and h_i. This implies that g_{i-1} and g_i are the same immersion. If $U_i \subset S - S'$, then $h_{i-1}^{-1}(U_i) \subset S - S'$ and the intermediate map between g_{i-1} and g_i fixes S'. If $\mathrm{cl}(U_i - S')$ is a closed disk, so is $\mathrm{cl}(h_{i-1}^{-1}(U_i) - S') = D_i'$. Moreover, g_{i-1} and g_i only differ in D_i'. This implies that the intermediate map between g_{i-1} and g_i can be taken fixing the subsurface S'. $\qquad\square$

Let $e : S \to S \times [-1, 1]$ be the embedding given by $e(x) = (x, 0)$, and let $p : S \times [-1, 1] \to S$ and $p' : S \times [-1, 1] \to [-1, 1]$ be the first and second

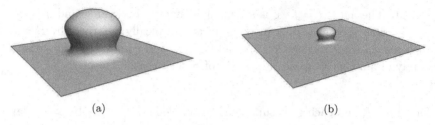

Fig. 6.6: A shrinking bump.

coordinate projections respectively. Let $H : S \times I \to S \times [-1,1]$ be an isotopy with $H_0 = e$. The map $p \circ H : S \times I \to S$ is a homotopy with $H_0 = \mathrm{id}_S$. By Lemma 6.11:

Lemma 6.13. *There exists $\varepsilon > 0$ such that $p \circ H_t : S \to S$ is a diffeomorphism for each $t \in [0, \varepsilon]$.* □

Taking $h = p \circ H$ and $z = p' \circ H$ we have $H_t(x) = (h_t(x), z_t(x))$. With this notation, Lemma 6.13 says that h is an ambient isotopy in S for a sufficiently small t.

Example 6.14. Consider the surface $S \times \{0\}$ with a "bump" such that the first coordinate projection is not one-to-one as in Fig. 6.6(a). Take a homotopy H that shrinks the bump homothetically until it disappears as $t \to 0$ (see Fig. 6.6(b)). There is no neighbourhood of 0 where $p \circ H_t : S \to S$ is one-to-one because H is not smooth.

Lemma 6.15. *There exists $\varepsilon > 0$ and a collection of isotopies*

$$H^1, \ldots, H^m : S \times [0, \varepsilon] \to S \times [-1, 1]$$

such that:

(1) $H_t^1 = e$ for all $t \in [0, \varepsilon]$;
(2) $H_0^1 = \cdots = H_0^m = e$;
(3) H_t^m and H_t are ambient isotopic in S, for all $t \in [0, \varepsilon]$; and
(4) there exists a pushing disk relating H_t^i and H_t^{i+1} for all $i = 1, \ldots, m-1$ and all $t \in (0, \varepsilon]$.

Moreover, if H fixes a compact subsurface $S' \subset S$, the maps $H^1, \ldots, H^m$ can be chosen fixing S'.

Proof. Fix $\varepsilon > 0$ satisfying the conditions of Lemma 6.13, and consider the horizontal and vertical components h and z of H as above.

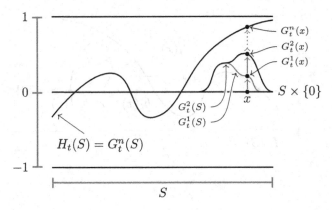

Fig. 6.7

For each $x \in S$ and $t \in [0, \varepsilon]$, the vertical segment $\{x\} \times [-1, 1]$ transversely meets the surface $H_t(S)$ at a single point. The height $z'_t(x) \in [-1, 1]$ at which $\{x\} \times [-1, 1]$ meets $H_t(S)$ is given by $z'_t(x) = z_t(h_t^{-1}(x))$. The last isotopy H^m of the collection we are looking for is given by the embeddings $H_t^m : S \to S \times [-1, 1]$ defined by

$$H_t^m(x) = \big(x, z'_t(x)\big),$$

for each $t \in [0, \varepsilon]$. That is, H^m and H are almost the same map: the images of H_t^m and H_t agree for all $t \in [0, \varepsilon]$, but H^m moves the points of $S \times \{0\}$ only in the vertical direction, correcting the horizontal movement induced on them by H. The ambient isotopy h of S satisfies that

$$H_t^m(h_t(x)) = \big(h_t(x), z_t(x)\big) = H_t(x).$$

This implies that the maps H_t^m and H_t are ambient isotopic in S for all $t \in [0, \varepsilon]$ (see Fig. 6.7).

Let $\mathcal{U}$ be a cover of S by open disks such that the closure of each element of $\mathcal{U}$ is a closed disk. Let $\Theta = \{\theta_1, \ldots, \theta_n : S \to I\}$ be a smooth partition of unity subordinate to $\mathcal{U}$. Since S is compact, Θ can be assumed to be finite. For each $j = 0, 1, \ldots, n$, we define an isotopy $G^j : S \times [0, \varepsilon] \to S \times [-1, 1]$ as follows (see Fig. 6.8):

$$G_t^0 = e;$$
$$G_t^j(x) = \big(x, (\theta_1 + \cdots + \theta_j)(x) \cdot z'_t(x)\big), \qquad \text{for } i = 1, \ldots, n.$$

Since $\theta_1 + \cdots + \theta_n \equiv 1$, the map G^n is precisely the isotopy H^m previously considered.

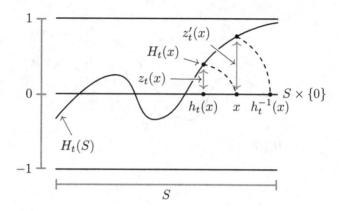

Fig. 6.8

Given $j = 1, \ldots, n$ and $t \in [0, \varepsilon]$, the embeddings G_t^{j-1} and G_t^j only differ over the disk $U_j \in \mathcal{U}$ which contains $\text{supp}(\theta_j)$. The closure $\text{cl}(U_j)$ of U_j is a closed disk in S whose image under G_t^{j-1} and G_t^j is contained in the interior of a 3-disk in $S \times [-1, 1]$. By Lemma 6.7 and the above discussion, the embeddings G_t^{j-1} and G_t^j are isotopic by two pushing disks. Call F_t^j the intermediate map between G_t^{j-1} and G_t^j. It is possible to choose them to depend smoothly on t. Since G_t^{j-1} and G_t^j approximates the embedding e as $t \to 0$, it can be assumed that F_t^j defines an isotopy $F^j : S \times [0, \varepsilon] \to S \times [-1, 1]$ such that $F_0^j = e$.

The maps

$$G^0, F^1, G^1, \ldots, G^{n-1}, F^n, G^n$$

verify the hypothesis of the statement.

Now, assume that H fixes a compact subsurface $S' \subset S$. Then, $z_t'(x) = 0$ for all $x \in S'$ and all $t \in [0, \varepsilon]$. As in the proof of Proposition 6.12, the cover $\mathcal{U}$ can be chosen so that each $U \in \mathcal{U}$ satisfies one of the following conditions:

(i) $U \subset S'$;
(ii) $U \subset S - S'$; or
(iii) $\text{cl}(U - S')$ is a closed 2-disk.

Assume that the two embeddings G_t^{j-1} and G_t^j only differ in the disk $U_j \in \mathcal{U}$. If $U_j \subset S'$, then G_t^{j-1} and G_t^j agree for each $t \in [0, \varepsilon]$ (in this case, the map F^j can be omitted). If $U_j \subset S - S'$, then G_t^{j-1} and G_t^j are isotopic by two pushing disks fixing S'. Finally, if $\text{cl}(U_j - S')$ is a closed 2-disk, then G_t^{j-1} and G_t^j only differ in the closed disk $\text{cl}(U_j - S')$ and, therefore, they are isotopic by two pushing disks fixing S'. Hence, it is possible to choose the isotopy F^j fixing S', for all $j = 1, \ldots, n$. □

Remark 6.16. It follows from the proof above that $H^1, \ldots, H^m$ can be chosen such that the pushed disks and the pushing balls relating H^i_t with H^{i+1}_t are "as small as we want", for any $t \in [0, \varepsilon]$ and $i = 1, \ldots, m - 1$. More precisely, given open covers $\mathcal{V}$ and $\mathcal{W}$ of S and $S \times [-1, 1]$ respectively, all the pushed disks and pushing balls can be assumed to be contained in elements of $\mathcal{V}$ and $\mathcal{W}$ respectively.

Theorem 6.17. *Let $f, g : S \to M$ be two regularly homotopic immersions. If there exists a regular homotopy H between f and g such that H_t is transverse for all $t \in I$, then f and g are regularly homotopic by transverse pushing disks. If in addition H fixes a compact subsurface $S' \subset S$, then f and g are regularly homotopic by transverse pushing disks fixing S'.*

Proof. Assume that the regular homotopy $H : S \times I \to M$ satisfies $H_0 = f$ and $H_1 = g$.

Since S and M are orientable, both f and g are 2-sided maps. Take a thickening $F : S \times [-1, 1] \to M$ of f. There is $\varepsilon > 0$ such that $H(S \times [0, \varepsilon])$ is contained in $F(S \times [-1, 1])$. Since F is a local diffeomorphism, there is a unique isotopy $\tilde{H} : S \times [0, \varepsilon] \to S \times [-1, 1]$ such that $F \circ \tilde{H} = H|_{S \times [0, \varepsilon]}$.

Making ε smaller if needed, there is a sequence of isotopies

$$\tilde{H}^1, \ldots, \tilde{H}^m : S \times [0, \varepsilon] \to S \times [-1, 1]$$

associated to $\tilde{H}$ as in Lemma 6.15. According to Remark 6.16, it is possible to assume that the restriction $F|_{B^i_t}$ is an embedding for all $t \in [0, \varepsilon]$ and all $i = 1, \ldots, m - 1$, where $B^i_t \subset S \times [-1, 1]$ is the pushing ball transforming $\tilde{H}^i_t$ into $\tilde{H}^{i+1}_t$.

For $i = 1, \ldots, m$, the map $H^i = F \circ \tilde{H}^i$ is a regular homotopy because F is an immersion. Moreover,

(i) $H^1_t = f$ for all $t \in [0, \varepsilon]$;

(ii) $H^1_0 = \cdots = H^m_0 = f$; and

(iii) $H^m_t = F \circ \tilde{H}^m_t$ and $H_t = F \circ \tilde{H}_t$ are ambient isotopic in S for all $t \in [0, \varepsilon]$.

Since the restriction of F to the pushing ball B^i_t connecting $\tilde{H}^i_t$ and $\tilde{H}^{i+1}_t$ is an embedding, its image $F(B^i_t)$ in M defines a pushing disk between H^i_t and H^{i+1}_t.

Taking ε even smaller if necessary, Lemma 6.11 implies that the immersions H^i_t are transverse for all $t \in [0, \varepsilon]$ and all $i = 1, \ldots, m$.

All this implies that H_t^i and H_t^{i+1} are transverse and they are related by a pushing disk, for all $i = 1, \ldots, m-1$ and all $t \in [0, \varepsilon]$. Hence, $H_t^1 = f$ and H_t^m are regularly homotopic by transverse pushing disks. On the other hand, since H_t^m and H_t are transverse and ambient isotopic in S, by Proposition 6.12, they are regularly homotopic by transverse pushing disks. Hence, f and H_t are regularly homotopic by transverse pushing disks, for any $t \in [0, \varepsilon]$.

If the regular homotopy H fixes a compact subsurface $S' \subset S$, then $\tilde{H}$ also fixes S'. According to Lemma 6.15, it can be assumed that $\tilde{H}^1, \ldots, \tilde{H}^m$ also fix S', and therefore $H_t^0 = f$ and H_t^m are regularly homotopic by transverse pushing disks fixing S' for all $t \in [0, \varepsilon]$. By Proposition 6.12 it turns out that the immersions f and H_t are regularly homotopic by transverse pushing disks fixing S', for all $t \in [0, \varepsilon]$.

In the same way, any $t \in I$ has an open neighbourhood $J_t \subset I$ such that H_t and $H_{t'}$ are regularly homotopic by transverse pushing disks (fixing S' if necessary) for all $t' \in J_t$. Then, the set

$$\left\{ t \in I \;\middle|\; \begin{array}{l} H_t \text{ and } f \text{ are regularly homotopic} \\ \text{by pushing disks (fixing } S') \end{array} \right\},$$

is a nonempty clopen set of I, hence it is I. Therefore f and $g = H_1$ are regularly homotopic by pushing disks (fixing S'). □

Theorem 6.18. *Let $f : S \to M$ be an immersion such that the points of M where f is not transverse are isolated. Let $H : S \times (-\varepsilon, \varepsilon) \to M$ be a regular homotopy such that $H_0 = f$ and H_t is transverse for all $t \in (-\varepsilon, \varepsilon)$, $t \neq 0$. Then, there exists $\varepsilon' < \varepsilon$ such that $H_{-\varepsilon'}$ and $H_{\varepsilon'}$ are regularly homotopic by transverse pushing disks. If H fixes a compact subsurface $S' \subset S$, then $H_{-\varepsilon'}$ and $H_{\varepsilon'}$ are regularly homotopic by transverse pushing disks fixing S'.*

Proof. Take a thickening $F : S \times [-1, 1] \to M$ of f. For a sufficiently small $\varepsilon_0 > 0$ it can be assumed that:

(i) $H(S \times (-\varepsilon_0, \varepsilon_0))$ is contained in $F(S \times [-1, 1])$;

(ii) there is an isotopy $\tilde{H} : S \times (-\varepsilon_0, \varepsilon_0) \to S \times [-1, 1]$ such that $\tilde{H}_0 = e$ and $F \circ \tilde{H} = H$; and

(iii) by Lemma 6.13 the map $p \circ \tilde{H}_t : S \to S$ is a diffeomorphism for all $t \in (-\varepsilon_0, \varepsilon_0)$, where $p : S \times [-1, 1] \to S$ is the first coordinate projection.

Assume that there is just one point $\bar{P} \in M$ where f is not transverse. By compactness of S, the preimage $f^{-1}(\bar{P})$ of $\bar{P}$ under f is a finite set

$\{P_1, \ldots, P_k\}$. Let B, B', B'' be three 3-disks in M such that

- $P \in \text{int}(B'') \subset B'' \subset \text{int}(B') \subset B' \subset \text{int}(B) \subset B$;
- $F^{-1}(B)$ is a disjoint union of k 3-disks in $S \times [-1, 1]$ such that F restricted to any of them is an embedding; and
- $f^{-1}(B)$ is a disjoint union of 2-disks in S such that f restricted to any of them is an embedding, and the same holds for $f^{-1}(B')$.

Consider

$$f^{-1}(B) = D_1 \cup \cdots \cup D_k \qquad f^{-1}(B') = D'_1 \cup \cdots \cup D'_k,$$

where D_i and D'_i are closed disks such that $P_i \in \text{int}(D'_i) \subset D'_i \subset \text{int}(D_i)$ for each $i = 1, \ldots, k$. For each $i = 1, \ldots, k$, define $U_i = \text{int}(D_i)$ and $U'_i = \text{int}(D'_i)$.

Consider an open cover $\mathcal{U}$ of S such that $U_1, \ldots, U_k \in \mathcal{U}$ and such that the only elements of $\mathcal{U}$ intersecting $U'_1 \cup \cdots \cup U'_k$ are $U_1, \ldots, U_k$. Let $\Theta = \{\theta_1, \ldots, \theta_n : S \to I\}$ be a smooth partition of unity subordinate to $\mathcal{U}$. Assume that $\theta_1, \ldots, \theta_m$, where $m < k$, are the elements of Θ whose support do not meet $U'_1 \cup \cdots \cup U'_k$, and set $\mu = \theta_1 + \cdots + \theta_m$. In particular $\mu(x) = 1$ for each $x \notin U_1 \cup \cdots \cup U_k$ and $\mu(x) = 0$ for each $x \in D'_1 \cup \cdots \cup D'_k$.

For a given $\varepsilon < \varepsilon_0$, define the homotopy $H^\varepsilon : S \times [-\varepsilon, \varepsilon] \to M$ by

$$H^\varepsilon(x, t) = H\big(x, \mu(x)(t + \varepsilon) - \varepsilon\big).$$

By compactness, H^ε_t uniformly converges to $H_0 = f$ as $(\varepsilon, t) \to (0, 0)$. Since f is transverse outside B'', there exists ε_1 with $0 < \varepsilon_1 < \varepsilon_0$ such that H^ε_t is transverse outside B'' for each $t \in [-\varepsilon_1, \varepsilon_1]$ (Lemma 6.11). Moreover, for each $\varepsilon < \varepsilon_0$ and $t \in [-\varepsilon, \varepsilon]$ the immersion H^ε_t agrees with $H_{-\varepsilon}$ over $D'_1 \cup \cdots \cup D'_k$, so it is transverse inside B''. By Theorem 6.17, we conclude that $H_{-\varepsilon_1} = H^{\varepsilon_1}_{-\varepsilon_1}$ and $H^{\varepsilon_1}_{\varepsilon_1}$ are regularly homotopic by transverse pushing disks leaving $D'_1 \cup \cdots \cup D'_k$ fixed.

The map $H^{\varepsilon_1}_{\varepsilon_1}$ agrees with H_{ε_1} outside $D_1 \cup \cdots \cup D_k$. By Lemma 6.7, modifying $H^{\varepsilon_1}_{\varepsilon_1}$ by two transverse pushing disks with pushed disk D_1, a transverse immersion that agrees with H_{ε_1} outside $D_2 \cup \cdots \cup D_k$ is obtained. By repeating this process with $D_2, \ldots, D_k$, we obtain a sequence of transverse pushing disks connecting $H^{\varepsilon_1}_{\varepsilon_1}$ with H_{ε_1}.

If H fixes the compact subsurface $S' \subset S$, at least one of the points $P_1, \ldots, P_k$ do not belong to S'. The balls B and B' of the previous construction can be chosen such that, for $i = 1, \ldots, k$, one of the following conditions holds:

(i) $D_i \subset S - S'$ (if $P_i \notin S'$);

(ii) $D_i \subset S'$ (if $P_i \in S' - \partial S'$); or

(iii) $D_i \cap S'$ and $D'_i \cap S'$ are closed disks (if $P_i \in \partial S'$).

With these assumptions the previous proof can be adapted to this case.

If there is more than one point of M were f is not transverse, the argument above can be applied in pairwise disjoint neighbourhoods of these points. $\qquad\square$

Now we are able to prove Theorem 6.6.

Proof of Theorem 6.6. (A) Let $H : S \times I \to M$ be a regular homotopy connecting f and g. For any $t \in I$, consider a thickening $F_t : S \times [-1, 1] \to M$ of H_t. For a sufficiently small $\varepsilon > 0$, there exists an isotopy

$$\tilde{H} : S \times [t - \varepsilon, t + \varepsilon] \to S \times [-1, 1]$$

such that $F_t \circ \tilde{H} = H|_{S \times [t-\varepsilon, t+\varepsilon]}$. Lemma 6.15 provides a neighbourhood $J_t \subset [t - \varepsilon, t + \varepsilon]$ of t where $\tilde{H}_t$ and $\tilde{H}_{t'}$ are isotopic by pushing disks for all $t' \in J_t$. Since F_t is a local diffeomorphism, making those pushing disks small enough (Remark 6.16), it can be assumed that the pushing balls are diffeomorphically mapped into M by F_t. Hence, the images by F_t of these pushing balls define a sequence of pushing disks connecting $\tilde{H}_t$ and $\tilde{H}_{t'}$. It follows that H_t and $H_{t'}$ are regularly homotopic by pushing disks for all $t' \in J_t$. By connectedness-based arguments similar to those at the end of the proof of Theorem 6.17, H_t and $H_{t'}$ are regularly homotopic by pushing disks for all $t, t' \in I$.

(B) It follows from Theorems 5.3 and 6.17, because the Homma-Nagase moves are particular cases of pushing disks.

(C) This point follows from Theorems 6.17 and 6.18. $\qquad\square$

Lemma 6.19. *If $f, g : S \to M$ are two regularly homotopic immersions and $D \subset S$ is a closed disk such that $f|_D$ and $g|_D$ are embeddings, there exists a regular homotopy with isolated singularities connecting f and g such that D is isotopically deformed by H.*

Proof. Let G be a regular homotopy with isolated singularities connecting f and g. By Lemma 6.8, there is a closed disk $D_1 \subset D$ such that D_1 is isotopically deformed by G.

Moreover, it is possible to construct an ambient isotopy η of S "shrinking" the disk D inwards until it becomes D_1 [30, p. 185]. Consider also η^{-1}, the ambient isotopy in S "opposite" to η defined by $\eta^{-1}(x, t) = \eta_t^{-1}(x)$, and put $G_1 = f \circ \eta$ and $G_2 = g \circ \eta^{-1}$. The concatenation $G_1 * G * G_2$ verifies the statement. $\square$

If $H : N \times I \to N'$ is a regular homotopy or an isotopy, the ambient isotopy $\bar{\zeta}$ of N' *extends* H if $H_t = \bar{\zeta}_t \circ H_0$ for all $t \in I$ or, equivalently, if the following diagram is commutative

$$
\begin{array}{ccc}
& N' \times I & \\
{\scriptstyle H_0 \times \mathrm{id}_I} \Big\uparrow & & \stackrel{\bar{\zeta}}{\searrow} \\
N \times I & \stackrel{H}{\longrightarrow} & N'
\end{array}
$$

where id_I denotes the identity map on I.

A key result in the theory of isotopies is the Isotopy Extension Theorem. See Theorem 1.3 of [30, p. 180] for a proof.

Theorem 6.20 (Isotopy Extension Theorem). *Let N and N' be two compact manifolds, and let $H : N \times I \to \mathrm{int}(N') \subset N'$ be an isotopy. Then, there exists an ambient isotopy $\bar{\zeta}$ of N' extending H.* $\square$

Lemma 6.21. *Let $f, g : \mathbb{S}^2 \to M$ be two regularly homotopic immersions such that $g(\mathbb{S}^2)$ is a 2-sphere standardly embedded into M. Assume that there exists a closed disk $D \subset \mathbb{S}^2$ such that f and g agree on D. Then, there exists a regular homotopy with isolated singularities $H : \mathbb{S}^2 \times I \to M$ connecting f and g that fixes D.*

Proof. By Lemma 6.19, there exists a regular homotopy with isolated singularities $F : \mathbb{S}^2 \times I \to M$ with $F_0 = f$ and $F_1 = g$, and such that D is isotopically deformed by F.

By the Isotopy Extension Theorem, there exists an ambient isotopy $\bar{\zeta}$ of M extending the isotopy $F|_{D \times I}$, that is, $\bar{\zeta}_t(F_0(x)) = F_t(x)$ for all $x \in D$ and $t \in I$. The map $F' : \mathbb{S}^2 \times I \to M$, given by

$$ F'_t(x) = \bar{\zeta}_t^{-1}(F_t(x)), $$

is a regular homotopy with isolated singularities fixing D and such that $F'_0(x) = f$.

The final immersion $g' = F'_1 = \bar{\zeta}_1^{-1} \circ g$ of the regular homotopy F' may not agree with the original immersion g. However, g' agrees with g (and f) on D, and $g'(\mathbb{S}^2) = \bar{\zeta}_1^{-1}(g(\mathbb{S}^2))$ is a 2-sphere standardly embedded in M. Since $\bar{\zeta}_1^{-1}$ is isotopic to the identity map of M, it preserves the orientation of M. By Palais' Disk Theorem (see [40, p. 52]), it is possible to construct an isotopy from g' to g which fixes D. The concatenation of this isotopy with F' is a regular homotopy with isolated singularities between f and g fixing D. □

The proof of Theorem 5.8 needs "well-behaved" pushing disks with respect to the pre-singular sets of the involved immersions. Let $f, g : S \to M$ be two transverse immersions related by the pushing disk (D_0, B_0). The pushing disk (D_0, B_0) is called *strongly transverse* if ∂D_0 in transverse to pre-Sing(f) and pre-Sing(g). Notice that the pre-singular sets of f and g agree on a small neighbourhood of ∂D_0 in $S - \text{int}(D_0)$. Moreover, ∂D_0 transversely intersects pre-Sing(f) at $P \in S$ if and only if it meets pre-Sing(g) transversely at P. If (D_0, B_0) is strongly transverse, the equator of B_0 contains neither triple points of f or g, nor tangencies with double curves of f or g.

Lemma 6.22. *If the immersions $f, g : S \to M$ are regularly homotopic by transverse pushing disks, they are regularly homotopic by strongly transverse pushing disks. If they are regularly homotopic by transverse pushing disks fixing a compact subsurface $S' \subset S$, then, for each compact subsurface $S'' \subset \text{int}(S')$, the immersions f and g are regularly homotopic by strongly transverse pushing disks fixing S''.*

Proof. Assume that f and g are related by the pushing disk (D_0, B_0).

If ∂D_0 does not intersect pre-Sing(f) or pre-Sing(g) transversely, there exists a closed disk D_1 with $D_0 \subset \text{int}(D_1)$ such that

(i) ∂D_1 transversely intersects pre-Sing(f); and
(ii) D_1 is close enough to D_0 to assume that both $f|_{D_1}$ and $g|_{D_1}$ are embeddings.

Since f and g agree on $S - D_0$, they also agree on $S - D_1$. By Lemma 6.7 there exist two transverse pushing disks (D_1, B_1) and (D_1, B_2) relating f and g, and since ∂D_1 transversely meets pre-Sing(f) and pre-Sing(g), the intermediate map between f and g can be taken such that the pushing disks (D_1, B_1) and (D_2, B_2) are strongly transverse.

If $S' \subset S$ is a compact subsurface such that f and g agree on S', then S' and D_0 are disjoint or they intersect at most at their boundaries. This implies that if $S'' \subset \text{int}(S')$ is a smaller subsurface, S'' and D_0 are disjoint, and therefore, D_1 can be chosen to be disjoint from S''. In this situation, f and g are related by two strongly transverse pushing disks fixing S''.

Now, if $f = f_0, f_1, \ldots, f_n = g : S \to M$ are immersions such that f_{i-1} and f_i are related by a transverse pushing disk (fixing S'), for all $i = 1, \ldots, n$. By the previous argument, f_{i-1} and f_i are related by strongly transverse pushing disks (fixing S') for all $i = 1, \ldots, n$. $\square$

Now, we are ready to state and prove our first Key Lemma.

Key Lemma 1. *Let $f, g : \mathbb{S}^2 \to M$ be two regularly homotopic immersions such that f is transverse and $g(\mathbb{S}^2)$ is a standardly embedded 2-sphere in M. Assume that there exists a closed disk $D \subset \mathbb{S}^2$ such that f and g agree on D, and let $D' \subset \mathrm{int}(D)$ be another closed disk. Then, f and g are regularly homotopic by strongly transverse pushing disks fixing D'.*

Proof. By Lemma 6.21, there exists a regular homotopy with isolated singularities between f and g fixing D. By Theorem 6.6, this implies that f and g are regularly homotopic by transverse pushing disks fixing D. By Lemma 6.22 f and g are regularly homotopic by strongly transverse pushing disks fixing D'. $\square$

From now on all the transverse pushing disks are assumed to be strongly transverse, unless otherwise stated

6.2 Shellings. Smooth triangulations

6.2.1 *Shellings*

Definition 6.23. Let N be a n-manifold with boundary. A closed n-disk $C \subset N$ is *free* in N if $C = N$ or if $C \cap \partial N$ is a closed $(n-1)$-disk.

Let B a closed n-disk and K be a regular cellular decomposition of B.

Definition 6.24. A sequence $\mathbf{s} = (C_1, \ldots, C_k)$ of different n-cells of K is a *partial shelling* of K if for each $i = 1, \ldots, k$, the set $\mathrm{cl}(C_i)$ is free in the closure of

$$B - \big(\mathrm{cl}(C_1) \cup \ldots \cup \mathrm{cl}(C_{i-1})\big).$$

A partial shelling $\mathbf{s}$ of K is a *shelling* of K if it contains all the n-cells of K. The cellular decomposition K is *shellable* if it has a shelling.

In the conditions of the above definition, if $\mathbf{s} = (C_1, \ldots, C_k)$ is a partial shelling of K, we say that it *starts* at C_1 and *ends* at C_k. Roughly speaking, a cellular decomposition of a closed n-disk is shellable if it is possible to remove its n-cells one by one while the remaining set at each step (if it is nonempty) is a closed n-ball. In the two-dimensional case, we have the next result, which follows from [60, Lemma 2].

Lemma 6.25. *If K is a regular cellular decomposition of a closed 2-disk and C is a 2-cell of K, there exists a shelling of K ending in C.* $\square$

While any cellular decomposition of a 2-disk is shellable, this is not true in higher dimensions. There exist unshellable cellular decompositions of the closed n-disk for each $n > 2$. However, any cellular decomposition of the n-disk has a shellable subdivision [9].

One of the main ingredients of the proof of Theorem 5.8 are the triangulations of a 3-manifold M, where the shellability of part (collections of possibly non-disjoint closed 3-balls) of these triangulations will be crucial. The work of Whitehead [19, 71] on simplicial complexes will ensure these conditions.

Let K be a simplicial complex, a simplex of it is *maximal* if it is not a proper face of another simplex. If ϵ^i is a maximal i-simplex of K, an $(i-1)$-face ϵ^{i-1} of it is *free* in K if it is not a face of a simplex of K other than ϵ^i. If ϵ^i is maximal in K and ϵ^{i-1} is an $(i-1)$-face of ϵ^i which is free in K, after removing both ϵ^i and ϵ^{i-1} from K, another simplicial complex K' is obtained. It is said that K' is obtained by *a collapsing of K* or by *collapsing ϵ^i from ϵ^{i-1}*. This is denoted by $K \searrow K'$, and the pair $(\epsilon^i, \epsilon^{i-1})$ is said to be an *i-collapsing* of K. The simplicial complex K *collapses* over a subcomplex K' if K' is obtained from K after a finite sequence of collapsings, and K is *collapsable* if it collapses to a point.

Theorem 6.26 ([71]). *If K is a finite simplicial complex there exists a stellar subdivision σK of K such that for each pair (B^n, B^{n-1}) σB^n collapses over σB^{n-1}, where B^n and B^{n-1} are simplicial n and $(n-1)$-balls of K respectively with $B^{n-1} \subset \partial B^n$.* $\square$

For triangulations of an n-ball, shellability implies collapsibility. The converse is not immediate because shellability requires that the space appearing at each step is an n-ball, while the space that appears after each collapsing might not even be a manifold. The example in [59] is an unshellable triangulation, which is in fact collapsible [14]. Anyway, the converse is "almost" true as the next result, which follows from [8, Theorem 6], shows:

Lemma 6.27. *If K is a collapsable triangulation of a closed 3-disk, then the second derived subdivision of K is shellable.* $\square$

The following results will also be helpful. The first one is Theorem 4 of [71], and the second one follows from the proof of Proposition 1 in [9].

Lemma 6.28. *Let K be a finite simplicial complex, and let σK be a stellar subdivision of K. If K collapses over a subcomplex L, then σK collapses over σL.* □

Lemma 6.29. *Let K be a triangulation of a closed n-disk, and let σK be a stellar subdivision of K. If K is shellable, σK is also shellable.* □

Let B a closed 3-disk. Given a triangulation T of B, define

$$N_T = \mathrm{cl}\Big(\bigcup_{v \in T^0 \cap \partial B} \mathrm{star}_T(v) \Big), \qquad C_T = \mathrm{cl}(B - N_T),$$

where T^0 is the 0-skeleton of T.

Lemma 6.30. *If T'' is the second derived subdivision of T, then $C_{T''}$ is a closed 3-disk and T'' collapses over $C_{T''}$.*

Proof. Since T'' is the second derived subdivision of T, $N_{T''}$ is a *collar* around ∂B, so $N_{T''}$ is homeomorphic to $\partial B \times I$, and $C_{T''}$ is a closed 3-disk.

On the other hand, since T'' is a second derived subdivision, it follows that for any simplex κ of T'' the intersection $\mathrm{cl}(\kappa) \cap \partial B$ is the closure of a face of κ.

Every triangle τ of T'' in ∂B is incident with a tetrahedron σ of T'', and it is possible to collapse σ from τ. So, collapse all the triangles of ∂B with their corresponding tetrahedra, and denote by B' the resulting simplicial complex.

If a is an edge of T'' in ∂B and a is incident with n tetrahedra of T'', then a is incident with $n - 2$ tetrahedra and $n - 1$ triangles contained in the interior of B, cyclically disposed around a. If $\sigma_1, \ldots, \sigma_{n-2}$ are those tetrahedra cyclically ordered around a, $\mathrm{cl}(\sigma_i) \cap \partial B = \mathrm{cl}(a)$ for all $i = 1, \ldots, n - 2$. Let $\tau_1, \ldots, \tau_{n-1}$ be the triangles of B' incident with a such that σ_i is incident with τ_i and τ_{i+1} for any $i = 1, \ldots, n - 2$. After the 3-collapsings $(\sigma_1, \tau_1), \ldots, (\sigma_{n-2}, \tau_{n-2})$, the triangle τ_{n-1} is the only one of B' incident with a that remains in the resulting complex. Hence, it is also possible to collapse (τ_{n-1}, a). After repeating the same operation on all the edges of T'' in ∂B a simplicial complex B'' is obtained.

Let v be a vertex T'' within ∂B. The structure of B'' around v is like a cone over $\mathrm{link}_{B''}(v)$. Moreover, if $\kappa \in B''$ is a simplex incident with v, then $\mathrm{cl}(\kappa) \cap \partial B = \{v\}$. The previous 2- and 3-collapsings performed on simplices of T'' incident with v have an associated sequence of 1- and 2-collapsings, respectively, in $\text{link}_{T''}(v)$. This sequence of collapsings define a collapsing of $\text{link}_{T''}(v)$ onto $\text{link}_{B''}(v)$. Since $\text{link}_{T''}(v)$ is a closed disk, it is collapsable, and therefore $\text{link}_{B''}(v)$ is also collapsable. Conversely, if we consider a collapsing of $\text{link}_{B''}(v)$ over any of its vertices, the associated sequences of 2- and 1-collapsings determine sequences of 3- and 2-collapsings respectively in the cone from v over $\text{link}_{B''}(v)$. After these sequences of collapsings the only simplex of T''' incident with v that remains in the resulting complex is the edge joining v with the last vertex of the collapsing of $\text{link}_{B''}(v)$. Collapsing this edge from v, all the simplices incident with v have been removed from B''. Since for each simplex $\kappa \in B''$ incident with v we have $\text{cl}(\kappa) \cap \partial B = \{v\}$, the same operation can be repeated for each vertex of T'' belonging to ∂B, obtaining in this way the required collapsing of B onto C_T. □

6.2.2   Smooth triangulations

The notion of *smooth triangulation* is introduced in [72] in order to relate the smooth and PL categories in the theory of manifolds. A triangulation of a n-manifold N is a homeomorphism $h : K \to N$, where K is a rectilinear simplicial complex in some euclidean space. If N has a differentiable structure, the triangulation h is *smooth* (with respect to that structure) if the restriction of h to each simplex of K is a diffeomorphism onto its image. Whitehead [72] proved that (see also [50]):

(i) any smooth manifold admits a smooth triangulation; and
(ii) any two smooth triangulations of a manifold have a common smooth subdivision.

Given a smooth triangulation T of the 3-manifold M, the map $f : S \to M$ is *simplicial with respect to T*, or that T *makes f simplicial*, if there exists a smooth triangulation K of S such that f is simplicial with respect to K and T. In our context, a key result is the following (for details and further related results see [66]):

Theorem 6.31. *If $f : S \to M$ is a transverse immersion of a surface S into a 3-manifold M, then there are smooth triangulations K and T of S and M, respectively, such that f is simplicial with respect to K and T.* □

From now on, all the considered triangulations are assumed to be smooth.

Let $f : S \to M$ be a filling immersion and set $\Sigma = f(S)$. If T is a triangulation of M that makes f simplicial, then Σ is a subcomplex of T and T induces a triangulation in the closure of each region of Σ. If R is a regular region of Σ, then T *shells* R if it induces a shellable triangulation of $\mathrm{cl}(R)$. If R is not a regular region, consider its characteristic map $\hat{q}_R : B^3 \to M$ (Sec. 1.6) and take the the the pull-back triangulation $\hat{T}$ of T by $\hat{q}_R$ in B^3. In this case, T *shells* R if $\hat{T}$ is shellable. The triangulation T *shells* f or Σ if T shells each region of $f(S)$.

Theorem 6.32. *Let $S_1, \ldots, S_k$ be a finite collection of surfaces, and let $f_i : S_i \to M$ be a transverse immersion for all $i = 1, \ldots, k$. Then, there exists a smooth triangulation T of M such that:*

(i) $f_1, \ldots, f_k$ are simplicial with respect to T;
(ii) if f_i is filling, then T shells f_i; and
(iii) if f_i and f_j are related by a pushing disk (D, B), the restriction of T to B collapses over $f_i(D)$ and over $f_j(D)$.

Proof. (i) By Theorem 6.31, for each $i = 1, \ldots, k$ there exists a triangulation T_i of M that makes f_i simplicial. By [72], $T_1, \ldots, T_k$ have a common subdivision T_0. Then, the maps $f_1, \ldots, f_k$ are all simplicial with respect to T_0. If T'_0 is a subdivision of T_0 in the conditions of Theorem 6.26, it is enough to choose T as the second derived subdivision of T'_0.

(iii) If f_i and f_j are related by a pushing disk (D, B), then B, $f_i(D)$ and $f_j(D)$ are simplicial with respect to T_0. Since T'_0 has been chosen following Theorem 6.26, the restriction of T'_0 to the 3-ball B collapses over $f_i(D)$ and over $f_j(D)$. Since collapsibility is preserved by stellar subdivisions [71] it is also preserved by derived subdivisions, and the restriction of T to B collapses over $f_i(D)$ and over $f_j(D)$.

(ii) If f_i is a filling immersion and R is a regular region of $f_i(S_i)$, then $\mathrm{cl}(R)$ is simplicial with respect to T_0. The election of T'_0 implies that the restriction of T'_0 to $\mathrm{cl}(R)$ is collapsable. By Lemma 6.27, the restriction of T to $\mathrm{cl}(R)$ is shellable. If R is not a regular region of f_i, let $\hat{q}_R : B^3 \to M$ be the characteristic map of R. The triangulations induced in $\mathrm{cl}(R)$ by T_0, T'_0 and T in $\mathrm{cl}(R)$ have their respective pull-back triangulations $\hat{T}_0$, $\hat{T}'_0$ and $\hat{T}$ of B^3 under $\hat{q}_R$. In particular, since T is the second derived subdivision of T'_0, then $\hat{T}$ is the second derived subdivision of $\hat{T}'_0$.

Taking $C_{\hat{T}} \subset B^3$ as in Lemma 6.30, $C_{\hat{T}}$ is a closed 3-disk and $\hat{T}$ collapses over $C_{\hat{T}}$. Since $C_{\hat{T}}$ is contained in $\mathrm{int}(B^3)$, its image under $\hat{q}_R$ is a closed 3-disk $\bar{C}$ contained in R. By Theorem 6.26, there is a stellar subdivision T'

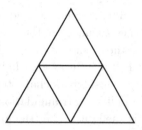

Fig. 6.9: A triangulation of $\mathbb{D}^2$ with no connected shelling.

of T whose restriction to $\bar{C}$ is collapsible. By Lemma 6.28, the pull-back triangulation $\hat{T}'$ of T' under $\hat{q}_R$ still collapses over $C_{\hat{T}}$. Moreover, the restriction of $\hat{T}'$ to $C_{\hat{T}}$ is collapsible, which implies that $\hat{T}'$ is collapsible. Denote by T'''' the second derived subdivision of T' in M. By Lemma 6.27, the pull-back triangulation $\hat{T}''''$ of T'''' in B^3 is shellable.

By Lemmas 6.28 and 6.29, T can be replaced by its stellar subdivision T''''. Repeating the same operation for all filling immersion f_i and all non-regular region of f_i, the required triangulation is obtained. $\square$

Definition 6.33. Under the hypothesis of the previous theorem, T is called a *good triangulation* of M with respect to $f_1, \ldots, f_k$.

Collapsibility and shellability are preserved by stellar subdivisions, hence:

Remark 6.34. If T is a good triangulation with respect to $f_1, \ldots, f_k$, so is any stellar subdivision of T.

6.2.3   Shellings of 2-disks

Let D be a closed 2-disk, T a triangulation of D and $\mathbf{s} = (t_1, \ldots, t_n)$ a shelling of T.

The shelling $\mathbf{s}$ is *connected* if for all $i = 2, \ldots, n-1$ the triangle t_i has at least one edge in $\mathrm{cl}(t_1 \cup \cdots \cup t_{i-1})$. While any triangulation of a two-dimensional disk has a shelling, in Fig. 6.9 it is depicted a simple triangulation with no connected shelling.

Take two simplicial arcs $\gamma, \delta \subset \partial D$ that intersect at most at their endpoints (in this case $\gamma \cup \delta = \partial D$). The shelling $\mathbf{s}$ *goes from γ to δ* if t_1 has an edge in γ and t_n has an edge in δ.

Proposition 6.35. *The second derived subdivision T'' of T has a connected shelling going from γ to δ, for any pair of paths as above.*

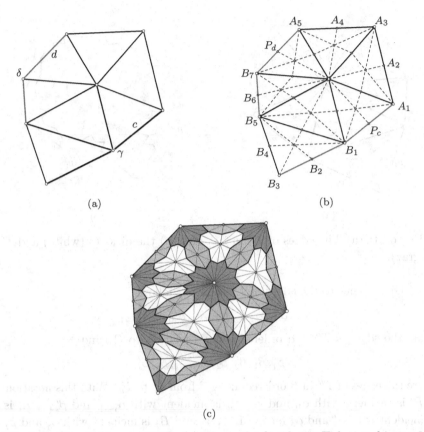

(a)

(b)

(c)

Fig. 6.10: Proof of Lemma 6.35: (a) The paths δ and γ in T depicted in gray. (b) The first derived subdivision T' of T with the vertices in ∂T labelled. (c) The stars in T'' of the vertices of T'.

Proof. Let T' be the first derived subdivision of T (Fig. 6.10(b)), and let $P'_1, \ldots, P'_m$ be the vertices of T'. The open stars $\operatorname{star}_{T''}(P'_1), \ldots, \operatorname{star}_{T''}(P'_m)$ of $P'_1, \ldots, P'_m$ in the triangulation T'' are pairwise disjoint, and they are the 2-cells of a cellular decomposition of D (Fig. 6.10(c)).

Let c and d be edges of γ and δ in T respectively (Fig. 6.10(a)). Let P_c and P_d be the two vertices of T' lying in c and d respectively (Fig. 6.10(b)). The points P_c and P_d are the common endpoints of two arcs α and β such that $\alpha \cup \beta = \partial D$. Let $A_1, \ldots, A_r$ $(B_1, \ldots, B_s)$ be the vertices of T' in α (β) ordered from P_c to P_d along α (β) (Fig. 6.11(a)).

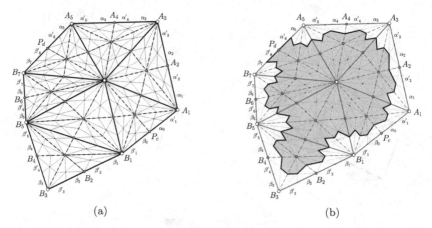

(a)   (b)

Fig. 6.11: (a) The edges of T'' in ∂T, and (b) the disks C (white) and E (gray).

Give names to the edges of T'' in ∂D as follows:

$$\alpha_0, \alpha_1', \alpha_1, \alpha_2', \alpha_2, \ldots, \alpha_r', \alpha_r, \alpha_{r+1}'$$

are the edges of T'' in α ordered along α from P_c to P_d, and

$$\beta_0, \beta_1', \beta_1, \beta_2', \beta_2, \ldots, \beta_s', \beta_s, \beta_{s+1}'$$

are the edges of T'' in β ordered along β from P_c to P_d. With this notation, P_c is incident with α_0 and β_0, P_d is incident with α_{r+1}' and β_{s+1}', A_i is incident with α_i' and α_i for $i = 1, \ldots, r$, and B_j is incident with β_j' and β_j for $j = 1, \ldots, s$ (Fig. 6.11(a)).

Consider the unique connected shelling $\mathrm{sh}(P_c)$ of the cell decomposition induced by T'' in $\mathrm{cl}(\mathrm{star}_{T''}(P_c))$ going from α_0 to β_0: $\mathrm{sh}(P_c)$ has all the triangles of $\mathrm{star}_{T''}(P_c)$ cyclically ordered from α_0 to β_0 around P_c, that is, the first triangle of the sequence $\mathrm{sh}(P_c)$ is incident with α_0 and the last one is incident with β_0.

Similarly, for each $i = 1, \ldots, r$, consider the connected shelling $\mathrm{sh}(A_i)$ of $\mathrm{cl}(\mathrm{star}_{T''}(A_i))$ going from α_i' to α_i, and for each $j = 1, \ldots, s$, take the connected shelling $\mathrm{sh}(B_j)$ of $\mathrm{cl}(\mathrm{star}_{T''}(B_j))$ going from β_j' to β_j.

The concatenation of shellings

$$\mathrm{sh}(\partial D) = \mathrm{sh}(P_c) * \mathrm{sh}(A_1) * \cdots * \mathrm{sh}(A_r) * \mathrm{sh}(B_1) * \cdots * \mathrm{sh}(B_s)$$

is a partial connected shelling of T'', and its first triangle contains an edge in γ. Define $C = \bigcup_{t'' \in \mathrm{sh}(\partial D)} \mathrm{cl}(t'')$ and $E = \mathrm{cl}(D - C)$. Both sets are closed disks (Fig. 6.11(b)).

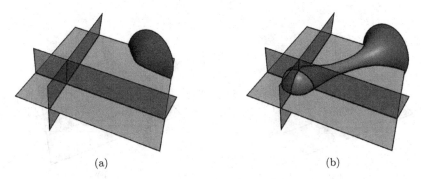

(a)   (b)

Fig. 6.12: Finger move $3/2$.

Let t''_B be a triangle of $\mathrm{star}_{T''}(P_d)$ with an edge in δ. By Lemma 6.25, the restriction of T'' to E has a shelling $\mathrm{sh}(E)$ ending at t''_B. The concatenation of shellings $\mathrm{sh}(\partial D) * \mathrm{sh}(E)$ is a shelling of D going from γ to δ. By construction, any triangle of T'' in E with an edge in ∂E has an edge in ∂C. If $\mathrm{sh}(E) = (\tau_1, \ldots, \tau_p)$, the triangle τ_k has at least an edge a in the boundary of $\mathrm{cl}(\tau_k \cup \cdots \cup \tau_p)$, for each $k = 1, \ldots, p$. Then, a either belongs to ∂E or is an edge of one of the triangles $\tau_1, \ldots, \tau_{k-1}$. If a belongs to ∂E, τ_k has at least an edge in ∂C. Therefore $\mathrm{sh}(\partial D) * \mathrm{sh}(E)$ is connected. $\square$

6.3 Complex f-moves

We will introduce some modifications only using f-moves applicable to filling Dehn surfaces. By Remarks 6.1 and 6.3, these modifications can be defined on Dehn surfaces or over their parametrizations. For the rest of the section, Σ denotes a filling Dehn surface of a 3-manifold M.

6.3.1 *Finger move $3/2$*

The modification of Σ depicted in Fig. 6.12 is called a *finger move $3/2$*. As for the rest of finger moves, it is a finger move $+3/2$ (resp. $-3/2$) if the deformation goes from left to right (resp. right to left) in the figure.

A finger move $+3/2$ can be decomposed into a finger move $+1$ (Fig. 6.13(a)) followed by a finger move $+2$ (Fig. 6.13(b)), therefore such a move is always filling preserving (Lemma 5.6). A finger move $-3/2$ can be obtained as a finger move -2 followed by a finger move -1. Hence, it is an f-move when the corresponding finger move -1 is an f-move.

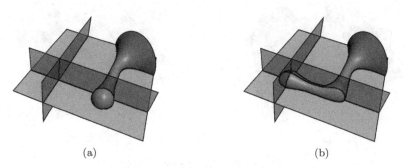

(a)   (b)

Fig. 6.13: Finger move $3/2$ decomposed into Haken moves.

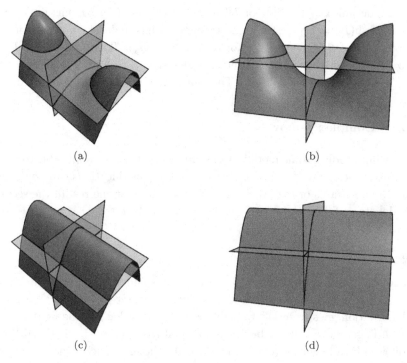

(a)   (b)

(c)   (d)

Fig. 6.14: Singular saddle move $+1$.

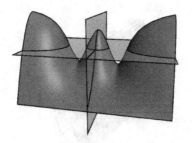

Fig. 6.15

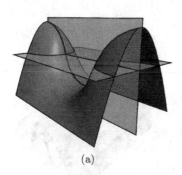

(a)

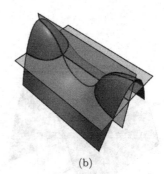

(b)

Fig. 6.16

6.3.2 Singular saddles

Consider now the *singular saddle moves* +1 *and* −1 depicted in Fig. 6.14 and Figs. 6.16 and 6.17 respectively. They are saddle moves where an additional sheet of Σ has been introduced.

Figures 6.14(a) and 6.14(b) show different viewpoints of the same scene. The same happens with Figs. 6.14(c) and 6.14(d). The deformation from Fig. 6.14(a) to Fig. 6.14(c) (resp. from Fig. 6.14(b) to Fig. 6.14(d)) is called a *singular saddle move* +1. By applying a finger move +1, Fig. 6.14(b) is deformed into Fig. 6.15. After that, Fig. 6.15 is transformed into Fig. 6.14(d) with two symmetric saddle moves. Both saddle moves are in the conditions of Proposition 5.11, so they are *f*-moves. Then, the singular saddle move +1 is always an *f*-move.

Figures 6.16(a) and 6.16(b) represent the same scene from slightly different points of view. The same happens with Fig. 6.18(a) and Fig. 6.18(b). After a *singular saddle move* −1, Fig. 6.17 is obtained. The transition

Fig. 6.17

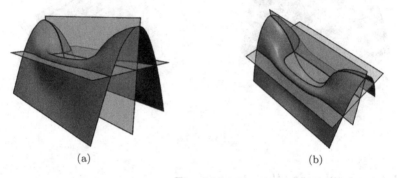

(a)   (b)

Fig. 6.18

Fig. 6.19

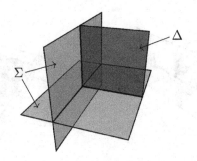

Fig. 6.20

from Fig. 6.16(a) to Fig. 6.18(a) (resp. from Fig. 6.16(b) to Fig. 6.18(b)) is made by a standard saddle move. Another saddle move produces Fig. 6.19. Finally, we pass from Fig. 6.19 to Fig. 6.17 by a finger move -1.

It is possible to show that the inverse construction to the one used to deform Fig. 6.17 to Fig. 6.16 is precisely a singular saddle move $+1$, i.e., singular saddle moves -1 and $+1$ are inverse moves of each other. A Dehn surface Σ' obtained from Σ after a singular saddle move -1 is a filling one if and only if the three Haken moves (two saddle moves plus finger move -1) that relate Σ with Σ' are f-moves.

6.3.3 *Pushing disks along a 2-cells*

Let R be a region of Σ. Consider a closed 2-disk Δ in M such that

(i) $(\partial \Delta, \text{int}(\Delta)) \subset (\partial R, R)$;
(ii) $\partial \Delta \cap \text{Sing}(\Sigma) \neq \emptyset$; and
(iii) Δ is transverse to Σ at $\partial \Delta$.

These conditions mean that the intersection of Δ with Σ at $\partial \Delta$ is as in Fig. 6.20 (see also Fig. 6.21). In particular, $\partial \Delta$ contains no triple points of Σ, and $\partial \Delta$ contains a finite number $n \geq 1$ of double points of Σ ($n = 4$ in Fig. 6.21). Notice that Δ is not an embedded closed disk in M, it is an embedded closed disk *with corners*, where the corners are the points of $\partial \Delta \cap S(\Sigma)$. In this situation, Δ is called an *n-gon in M outside* Σ. The *vertices* of Δ are the points of $\text{Sing}(\Sigma) \cap \partial \Delta$, and the *edges* of Δ are the connected components of $\partial \Delta - \{\text{vertices of } \Delta\}$.

Let Δ be an n-gon outside Σ with $n \geq 2$. If α is an edge of Δ, the sheet of Σ containing α can be modified in the way shown in Figs. 6.21 and 6.22. Figures 6.21(a) and 6.21(b) represent the same scene from different points

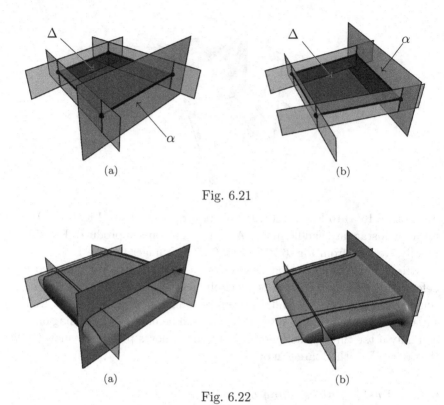

Fig. 6.21

Fig. 6.22

of view, and the same occurs with Figs. 6.22(a) and 6.22(b). After the modification, Fig. 6.21(a) is transformed into Fig. 6.22(a) (and Fig. 6.21(b) into Fig. 6.22(b)). It is said that the resulting Dehn surface Σ' has been obtained *pushing Σ from α along Δ*, or that Σ' is obtained from Σ *by a pushing disk along Δ from α*.

It is possible to define the same operation starting with a n-gon contained in Σ, that is, with Δ the closure of a regular face of Σ (Sec. 2.3). Now, the vertices of Δ are the triple points of Σ that lay on $\partial\Delta$, and the edges of Δ are the edges of Σ contained in $\partial\Delta$. In this case, Δ is said to be an n-*gon inside* Σ. If Δ is an n-gon inside Σ with $n \geq 2$ and α is an edge of Δ, it is possible to push Σ along Δ from α as in the outside-Σ case (just imagine that in Fig. 6.21 there is a horizontal sheet of Σ containing Δ), and as above, it is said that the new Dehn surface Σ' is obtained *pushing Σ from α along Δ*, or that Σ' is obtained from Σ *by a pushing disk along Δ from α*.

Assume that Σ' is obtained from Σ by a pushing disk along an n-gon Δ inside or outside Σ. Let $f : S \to M$ be a parametrization of Σ and let g be

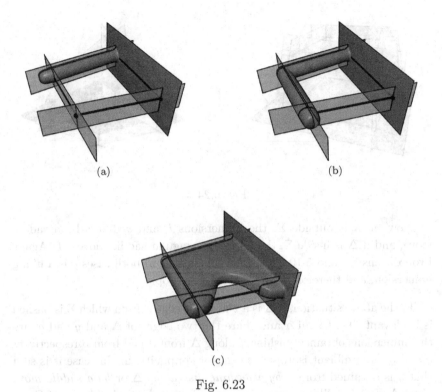

(a)   (b)

(c)

Fig. 6.23

the parametrization of Σ' obtained from f after the deformation.

Proposition 6.36. *If f and g are filling, they are $\boldsymbol{f}$-homotopic. Moreover, if f is filling and Δ has more than two edges, g is filling.*

Proof. Assume that f and g are filling immersions and that $n = 2$. If Δ is outside Σ, the pushing disk along Δ is just a saddle move. Therefore, by Definition 5.7, f and g are $\boldsymbol{f}$-homotopic. If Δ is inside Σ, the pushing disk along Δ is a singular saddle move -1, which is an $\boldsymbol{f}$-move because both f and g are filling. By Sec. 6.3.2, this implies that f and g are $\boldsymbol{f}$-homotopic.

Assume now that $n > 2$ and that f is a filling immersion. By applying $n - 2$ consecutive finger moves to the immersion f, a situation like the one depicted in Fig. 6.23(b) is obtained. Then, after an ambient isotopy the immersion h of Fig. 6.23(c) is obtained. If Δ is outside Σ, all those finger moves are finger moves $+1$. If Δ is inside Σ, all of them are finger moves $+2$. In either case, the immersion h is filling.

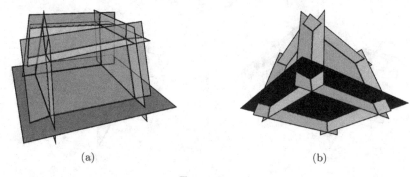

(a)   (b)

Fig. 6.24

Now, if Δ is outside Σ, the immersions h and g differs by a saddle move, and if Δ is inside Σ, they differ by singular saddle move -1. Again, Propositions 5.9 and 5.10 allow to conclude that in both cases g is a filling immersion, and therefore h and g are f-homotopic. $\square$

In the above situation, if Δ is a 2-gon, the edge α from which Σ is pushed is irrelevant. That is, if α and β are the two edges of Δ and g and g' are the immersions obtained pushing f along Δ from α and from β respectively, g and g' are ambient isotopic. Then, for simplicity, in this case it is said that g is obtained from f *by a pushing disk along Δ* or *by a saddle move along Δ*. If f is a filling immersion and g is not, Δ is a *bad 2-gon* of Σ.

6.3.4 Pushing disks along 3-cells

Let R_0 be a regular region of Σ such that all its faces are also regular (as in Fig. 6.24). Such a region is called *∂-regular*.

Let Δ_0 be a face of R_0 (as the 4-gon of the dark horizontal sheet in Fig. 6.24(a)). The ∂-regularity of R_0 implies that $\mathrm{cl}(\Delta_0)$ is a closed 2-disk and that $\mathrm{cl}(R_0)$ is a closed 3-disk. Consider a parametrization $f : S \to M$ of Σ, and the immersion $g : S \to M$ obtained from f by a pushing disk (D_0, B_0) as in Figs. 6.24 and 6.25. Figures 6.24(a) and 6.24(b) represent the same scene from different viewpoints. The same happens with Figs. 6.25(a) and 6.25(b).

The pushing disk (D_0, B_0) transforms Fig. 6.24(a) (Fig. 6.24(b)) into Fig. 6.25(a) (Fig. 6.25(b)). There exists an open disk $\tilde{\Delta}_0$ in S such that the restriction of f to $\tilde{\Delta}_0$ is an embedding and $f(\tilde{\Delta}_0) = \Delta_0$. The interior of the pushed disk D_0 in S contains $\mathrm{cl}(\tilde{\Delta}_0)$, and it is as close to $\tilde{\Delta}_0$ as needed to guarantee that $f|_{D_0}$ is an embedding. The pushing ball B_0 contains the

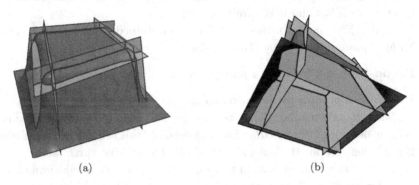

(a) (b)

Fig. 6.25

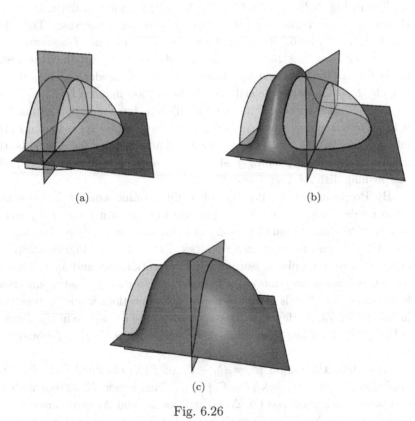

(a) (b)

(c)

Fig. 6.26

region R_0 and $\mathrm{cl}(R_0) \cap \partial B_0 = \mathrm{cl}(\Delta_0)$. The disk $g(D_0)$ is located in the exterior of $\mathrm{cl}(R_0)$ and it is parallel to $\partial R_0 - \Delta_0$.

Define $\Sigma' = g(S)$. In this situation g is obtained from f (and Σ' from Σ) *by a pushing disk along R_0 from Δ_0.*

Proposition 6.37. *If g is a filling immersion, it is f-homotopic to f.*

Proof. The proof is by induction on the number m of faces of R_0.

Since R_0 is ∂-regular, the vertices, edges and faces of R_0 define a regular cellular decomposition of ∂R_0. The 1-skeleton of this cellular decomposition is a 3-valent graph. By Euler's formula, R_0 has at least three faces.

Assume first that $m = 3$. In this case, R_0 can only be a trihedron or a 1-gonal prism. Since R_0 is ∂-regular, it must be a trihedron like in Fig. 6.26(a). A saddle move transforms Fig. 6.26(a) into Fig. 6.26(b), and the deformation applied in Fig. 6.26(c) to obtain Fig. 6.26(b) is indeed a finger move $+1$. Hence, the immersion h of Fig. 6.26(b) is a filling immersion. Thus, the saddle move of Fig. 6.26(b) is an f-move and f and g are f-homotopic.

Assume now that R_0 has m faces, with $m > 3$, and that the result holds for any region with less than m faces. Consider the closed disk $\Gamma = \mathrm{cl}(\partial R_0 - \Delta_0)$ with the cellular decomposition induced by Σ.

Since R_0 is ∂-regular, Γ has no self-adjacent faces and, by Lemma 6.25, it has a free face Δ_1. Since Δ_0 and Δ_1 are regular, and Δ_1 is free in Γ, the intersection $\mathrm{cl}(\Delta_0) \cap \mathrm{cl}(\Delta_1)$ is the closure of an edge α of Δ_0. Let f' be the immersion obtained pushing f along Δ_1 from α, and denote by (D_1, B_1) the pushing disk relating f and f'.

By Proposition 6.36, if f' is not a filling immersion, $\mathrm{cl}(\Delta_1)$ is a bad 2-gon inside Σ. By Sec. 6.3.2, it is possible to transform f into f' by means of two saddle moves parallel to Δ_1 at both sides of Δ_1 (see Figs. 6.28(a) and 6.28(b)) and a finger move -1 (see Fig. 6.28(c)). Figure 6.28(d) is obtained after an ambient isotopy. By Propositions 5.9 and 5.10, if any of these moves is not an f-move it is because there are some "bad" connections between 1-, 2- or 3-cells of Σ, but these bad connections would be preserved in the immersion g. Hence, if Δ_1 is a bad 2-gon, g cannot be filling. Since g is filling, f' is a filling immersion. By Proposition 6.36, f' is f-homotopic to g.

Now, take the region $R_0' = R_0 - B_1$ of $f'(S)$ obtained from R_0 after applying the pushing disk (D_1, B_1) to f. This region R_0' agrees with R_0 except on a neighbourhood of Δ_1. The faces Δ_0 and Δ_1 of R_0 are replaced by a single face Δ_0' of R_0'. Since $\mathrm{cl}(\Delta_0)$ and $\mathrm{cl}(\Delta_1)$ meet only along $\mathrm{cl}(\alpha)$ and R_0 is ∂-regular, R_0' is also ∂-regular and it has $m - 1$ faces. The induction

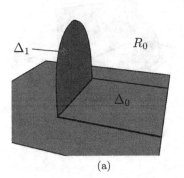 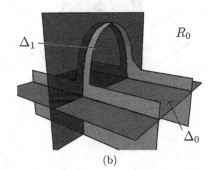

Fig. 6.27

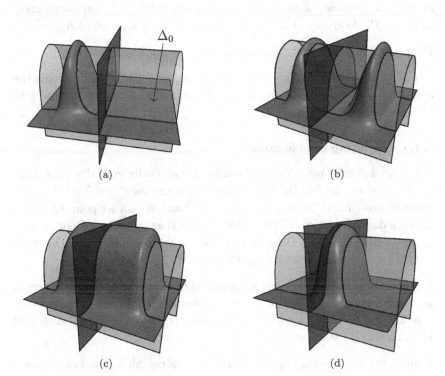

Fig. 6.28

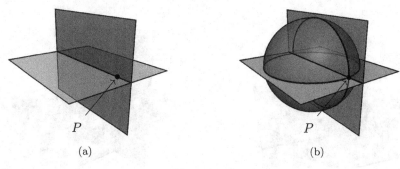

(a) (b)

Fig. 6.29

hypothesis implies that f' and g are f-homotopic.   □

The ∂-regular region R_0 is *superregular* if there are no cells of Σ having multiple adjacencies with R_0, that is, for any cell ϵ of Σ the intersection $\mathrm{cl}(\epsilon) \cap \mathrm{cl}(R_0)$ is empty or it is the closure of exactly one cell of R_0.

Proposition 6.38. *If R_0 is superregular, g is a filling immersion.*

Proof. The proof is straightforward. The superregularity of R_0 avoids the appearance of bad connections among the cells of Σ' that might cause g to be non-filling.   □

6.3.5   Inflating double points

Let P be a double point of Σ. Consider a standardly embedded 2-sphere Σ_P in M as in Fig. 6.29(b). The sphere Σ_P contains P and $\Sigma_P \cap \Sigma$ is the union of two circles. These circles meet in P and in another point Q, which is also a double point of Σ. The union $\Sigma \cup \Sigma_P$ is again a filling Dehn surface of M. Consider a filling Dehn surface $\Sigma \# \Sigma_P$ obtained connecting Σ_P with Σ by a spiral piping around P (see Sec. 2.4.5).

Proposition 6.39. *The spiral piping can be chosen such that every parametrization $f : S \to M$ of Σ is f-homotopic to a parametrization g of $\Sigma \# \Sigma_P$ that agrees with f except on a closed disk $D \subset S$ such that $D \cap \mathrm{pre\text{-}Sing}(f)$ is an arc.*

Proof. We start modifying the surface Σ in a neighbourhood of P which is small enough in comparison with the 2-sphere Σ_P (see Fig. 6.30(a)). Figure 6.30(c) shows Σ modified by two consecutive finger moves $+1$ (see also Fig. 6.30(b)).

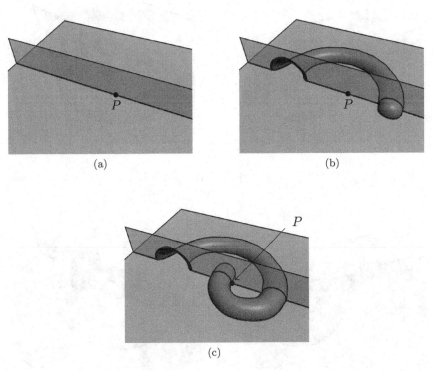

(a)

(b)

(c)

Fig. 6.30

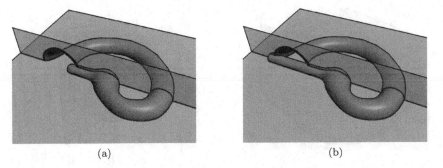

(a)

(b)

Fig. 6.31

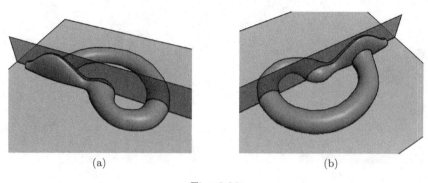

(a)   (b)

Fig. 6.32

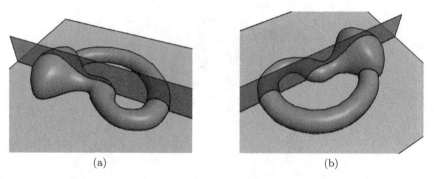

(a)   (b)

Fig. 6.33

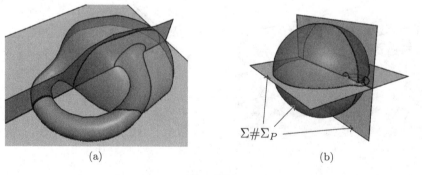

$\Sigma \# \Sigma_P$

(a)   (b)

Fig. 6.34

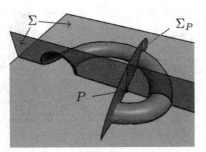

Fig. 6.35

After an ambient isotopy Fig. 6.31(a) is obtained. Another finger move $+1$ produces the situation depicted in Fig. 6.31(b).

Now "close the tunnel entrance" with two saddle moves, one above and the other below the sheet of Σ depicted horizontally in Fig. 6.31(b). The resulting Dehn surface is depicted in Fig. 6.32(a). Proposition 5.11 implies that these two saddle moves are f-moves and hence, the last surface is filling. Figure 6.32(b) shows the same surface from a different point of views. By ambient isotopy Fig. 6.32(a) (resp. Fig. 6.32(b)) is transformed into Fig. 6.33(a) (resp. Fig. 6.33(b)). Finally, "inflate the balloon" to obtain the Dehn surface of Fig. 6.34(a), which is obviously ambient isotopic to Fig. 6.34(b).

The surface $\Sigma \# \Sigma_P$ obtained after this process is identical to $\Sigma \cup \Sigma_P$ except in a little neighbourhood of P, where it looks like in Fig. 6.35. By construction, if $f : S \to M$ is a parametrization of Σ, all these modifications transform f into a parametrization g of $\Sigma \# \Sigma_P$ that agrees with f except in the pushed disk of the first finger move $+1$ of Fig. 6.30(b), which is the piping disk in the domain of Σ of the spiral piping that connects Σ with Σ_P. $\qquad\square$

The Dehn surface $\Sigma \# \Sigma_P$ of Proposition 6.39 is said to be obtained from Σ after *inflating* P. A similar notation will be used for parametrizations.

6.3.6 *Passing through spiral pipings*

Some of the good properties of spiral pipings have already been discussed in Sec. 2.4.5.

Another good property is that *it does not disturb f-moves* in the following sense. Let $f, g : S \to M$ be two filling immersions such that g is obtained from f by a finger move $+2$ through a triple point P of f. Let S', f' and g'

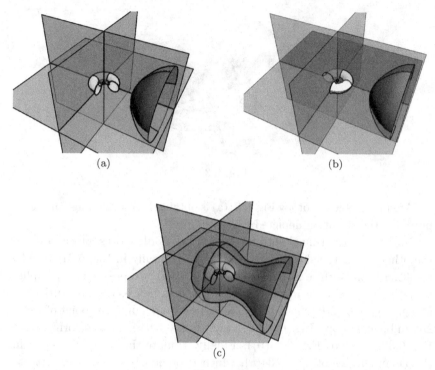

(a)   (b)

(c)

Fig. 6.36

be the surface and the immersions obtained from S, f and g respectively after a spiral piping around P. Assume that the spiral piping is small enough with respect to the finger move $+2$ so the situation is like the one depicted in Fig. 6.36.

Lemma 6.40. *The filling immersions f' and g' are f-homotopic.*

Proof. The spiral piping can be in two different relative positions with respect to the finger move $+2$ (Figs. 6.36(a) and 6.36(b)). Assume that it is as in Fig. 6.36(a). The proof of the other case is similar.

After an ambient isotopy (Fig. 6.37(a)), and a finger move $+1$ followed by two f-saddle moves, one above and the other below the horizontal sheet, "the tunnel entrance is closed" in the same way as in Figs. 6.31 and 6.32, and Fig. 6.37(c) is obtained. Figure 6.37(d) shows the same scene as Fig. 6.37(c) from a different viewpoint. The Dehn surface (still denoted by) Σ forms, with the spiral piping, a configuration of a "bisected cylinder" as in Fig. 6.38(a). It is possible to push Σ along this cylinder using two pushing disks along

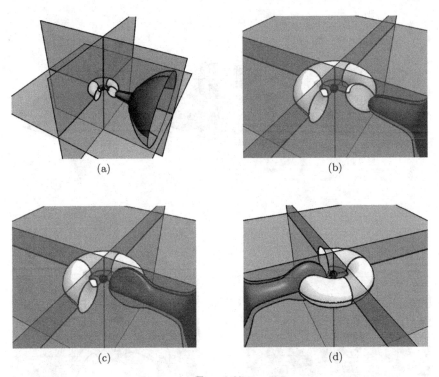

(a)

(b)

(c)

(d)

Fig. 6.37

3-cells: a first pushing disk along 2-gonal prism (Fig. 6.38(b)) and a second pushing disk along a trihedron (Fig. 6.38(c)). The Dehn surfaces obtained after each of these pushing disks are filling, and by Proposition 6.37 they are f-homotopic to the original surface Σ. Figure 6.39(a) is obtained after applying this operation twice. Figure 6.39(b) is the same scene seen from a different point of view. Figure 6.39(b) is transformed into Fig. 6.39(c) by a pushing disk along of a 3-gon inside Σ. By Proposition 6.36, this operation can be decomposed into f-moves. After that, a filling-preserving singular saddle move -1 leads to Fig. 6.39(d). Figure 6.39(e) shows (from other point of view) the resulting configuration after these operations and and ambient isotopy, and Fig. 6.39(f) displays the Dehn surface after another singular saddle move -1 (which also is an f-move). Finally, by ambient isotopy (see Figs. 6.39(g) to 6.39(j)) the required surface is obtained (Fig. 6.36(c)). □

Under the hypothesis of this lemma, the immersion g' is said to be *obtained from f' by a piping passing move through P.*

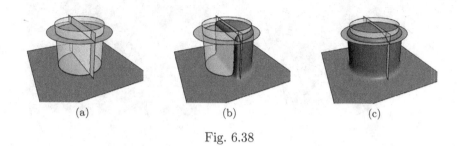

(a)   (b)   (c)

Fig. 6.38

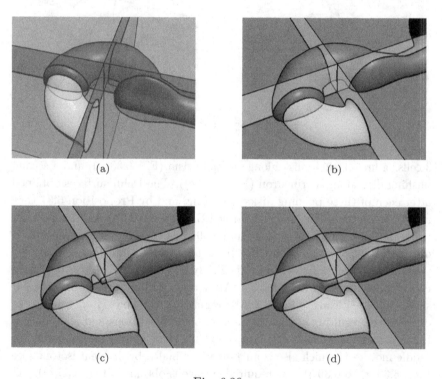

(a)   (b)

(c)   (d)

Fig. 6.39

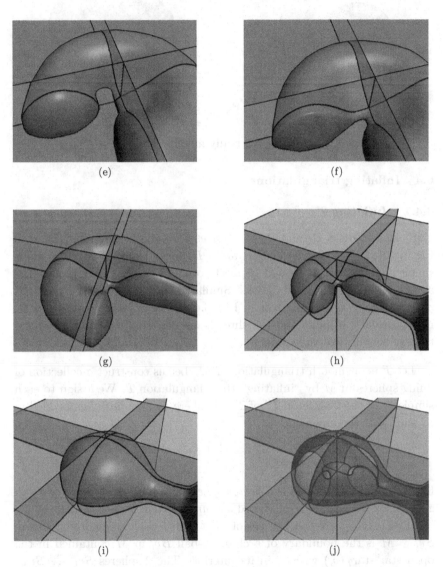

Fig. 6.39: (cont.)

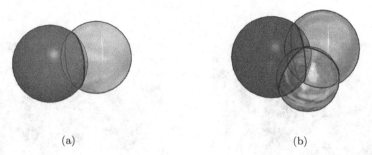

(a)   (b)

Fig. 6.40: Normally meeting balls.

6.4   Inflating triangulations

6.4.1   *Inflating T*

Two closed 3-disks B_1 and B_2 of M *meet normally* if they intersect as in Fig. 6.40(a). The 2-spheres ∂B_1 and ∂B_2 meet transversely in a single simple double curve. If B_1 and B_2 meet normally, then $B_1 \cap B_2$, $\mathrm{cl}(B_1 - B_2)$ and $\mathrm{cl}(B_2 - B_1)$ are closed 3-disks. Similarly, three 3-disks B_1, B_2 and B_3 *meet normally* if they meet as in Fig. 6.40(b). The balls B_1, B_2 and B_3 meet normally in pairs and the three 2-spheres ∂B_1, ∂B_2 and ∂B_3 meet transversely in two triple points.

Let T be a smooth triangulation of M. Let us construct a collection of filling spheres in M by "inflating" the triangulation T. We assign to each simplex ϵ of the 2-skeleton T^2 of T a 2-sphere $S\epsilon$ standardly embedded in M such that the union

$$\Sigma_T = \bigcup_{\epsilon \in T^2} S\epsilon$$

fills M. The 2-sphere $S\epsilon$ is a *vertex sphere* of Σ_T if ϵ is a vertex of T, an *edge sphere* if ϵ is an edge of T, and a *triangle sphere* if ϵ is a triangle of T.

If $v_1, \ldots, v_{m_0}$ are the vertices of M, for all $i = 1, \ldots, m_0$, the 2-sphere $Sv_i \subset M$ is the boundary of a closed 3-ball Bv_i in M contained in the open star $\mathrm{star}_T(v_i)$ with v_i in its interior. The 2-spheres $Sv_1, \ldots, Sv_{m_0}$ are all disjoint (Figs. 6.41(a) and 6.41(b)). For each $i = 1, \ldots, m_0$ the triangulation T induces a triangulation T_{Sv_i} of Sv_i which is a projection of that of $\mathrm{link}_T(v_i)$ (Figs. 6.41(c) and 6.41(d)): every i-simplex $\epsilon^i \in \mathrm{star}_T(v_i)$ meets Sv_i transversely, and $\epsilon^i \cap Sv_i$ is an $(i-1)$-simplex of T_{Sv_i}.

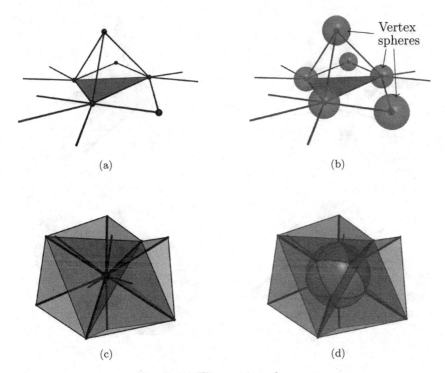

(a) (b)

(c) (d)

Fig. 6.41: The vertex spheres.

If $e_1, \ldots, e_{m_1}$ are the edges of M, for each $j = 1, \ldots, m_1$, the 2-sphere $Se_j \subset M$ is the boundary of a closed 3-disk $Be_j \subset M$ as in Fig. 6.42(b). The 3-disk Be_j is contained in the open star $\mathrm{star}_T(e_j)$ and meets e_j in a closed arc $\tilde{e}_j \subset e_j$. The 2-sphere Se_j and the edge e_j meet transversely at the endpoints of $\tilde{e}_j$.

Choose $Be_1, \ldots, Be_{m_1}$ to be pairwise disjoint (Fig. 6.42(a)), and such that for each $i \in \{1, \ldots, m_0\}$ and each $j \in \{1, \ldots, m_1\}$ the 3-disks Bv_i and Be_j are disjoint unless v_i and e_j are incident. In this case, the 3-disks Bv_i and Be_j meet normally and $Bv_i \cap Be_j \cap e_j$ is another sub-arc of e_j. If the points of $Se_j \cap e_j$ are regarded as the "poles" of Se_j, each triangle t of M incident with e_j meets Se_j transversely in an open arc which is the interior of a "meridian" a connecting the poles. The intersection $\mathrm{cl}(t) \cap Be_j$ is a closed disk whose boundary is $a \cup \tilde{e}_j$.

Finally, if $t_1, \ldots, t_{m_2}$ are the triangles of M, for each $k = 1, \ldots, m_2$ the 2-sphere St_k is the boundary of a 3-disk Bt_k as in Fig. 6.43(b). The 3-disk Bt_k is contained in $\mathrm{star}_T(t_k)$, and it intersects t_k in a closed disk $\tilde{t}_k \subset t_k$.

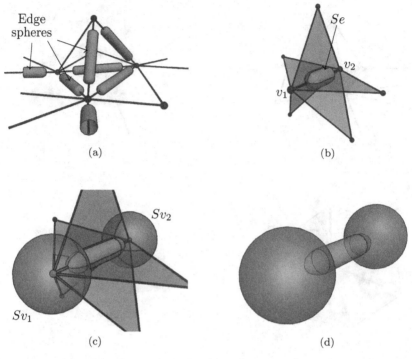

Fig. 6.42

The intersection of St_k and t_k is transverse. For each $\epsilon \in T^2$ (other than t_k), the 3-disks Bt_k and $B\epsilon$ are disjoint unless ϵ is incident with t_k. In this case, Bt_k and $B\epsilon$ meet normally. Moreover, if $v_i < e_j < t_k$, the 3-disks Bv_i, Be_j and Bt_k meet normally (see Figs. 6.43(c), 6.43(d) and 6.46) and each triple point of $Sv_i \cap Se_j \cap St_k$ is contained in one of the two tetrahedra of T incident with t_k.

Figure 6.44 shows how the 2-spheres of Σ_T meet inside of a 3-disk Bv for some vertex $v \in T$.

This construction can be easily generalized to the case where T is a cellular decomposition of M instead of a triangulation.

If K is any subset of simplices of T^2 (a subcomplex, for instance), define

$$\Sigma_K = \bigcup_{\epsilon \in K} S\epsilon.$$

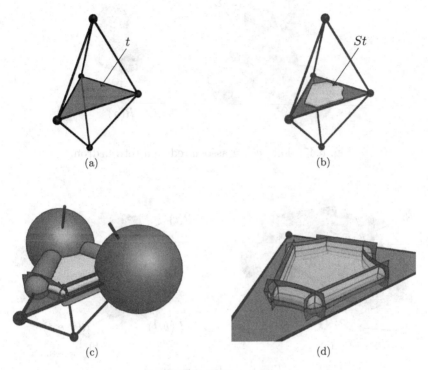

(a)

(b)

(c)

(d)

Fig. 6.43

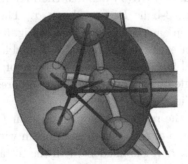

Fig. 6.44

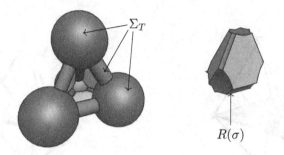

Fig. 6.45: The region associated to a tetrahedron.

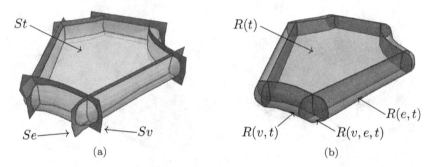

Fig. 6.46: Region names.

6.4.2   *Naming the regions of* Σ_T

If σ is a tetrahedron of T, $R(\sigma)$ denotes the region of Σ_T obtained after removing from σ the 3-balls $B\epsilon$ for all $\epsilon < \sigma$. The closure of $R(\sigma)$ is a "truncated tetrahedron" as the one depicted in Fig. 6.45 with a 6-gonal face in the 2-sphere St for each triangle t of σ, a 4-gonal face in Se for each edge e of σ, and a 6-gonal face in Sv for each vertex v of σ.

Let t be a triangle of T. The remaining 2-spheres of Σ_T meet St as in Fig. 6.46 (see also Fig. 6.43).

Let $R(t)$ denote the region of Σ_T contained in Bt and in no other $B\epsilon$ with $\epsilon \lneq t$. The closure of $R(t)$ is a 6-gonal prism with a 4-gonal face in Se for all $\epsilon \in T$ with $\epsilon \lneq t$. If t is incident with the tetrahedra σ_1 and σ_2 of T, the two 6-gonal faces of $R(t)$ are also faces of $R(\sigma_1)$ and $R(\sigma_2)$ respectively.

If e is an edge of T incident with t and v_1, v_2 are the vertices of e, then $R(e, t)$ denotes the region of Σ_T contained in $Be \cap Bt$ and not contained in

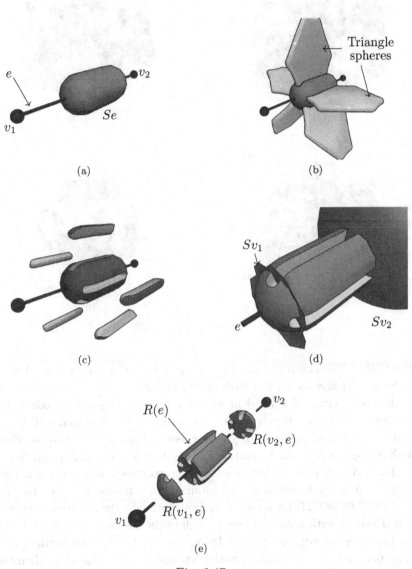

Fig. 6.47

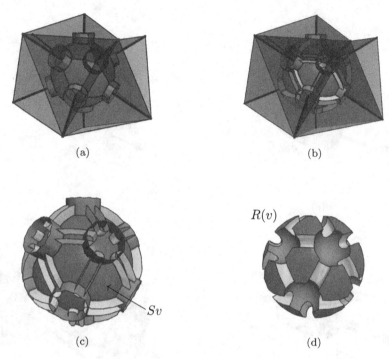

(a)

(b)

$R(v)$

Sv

(c)

(d)

Fig. 6.48

$Bv_1 \cup Bv_2$. The closure of $R(e,t)$ is a 2-gonal prism with its 4-gonal faces in Se and St and its 2-gonal faces in Sv_1 and Sv_2.

If v is a vertex of T incident with t and, e_1 and e_2 are the edges of t incident with v, then $R(v,t)$ denotes the region of Σ_T contained in $Bv \cap Bt$ and not contained in $Be_1 \cup Be_2$. The closure of $R(v,t)$ is a 2-gonal prism, with its 4-gonal faces in Sv and St and its 2-gonal faces in Se_1 and Se_2.

If v and e are a vertex and an edge of T respectively with $v < e < t$, then $R(v,e,t)$ denotes the region of Σ_T contained in the intersection $Bv \cap Be \cap Bt$. Since Bv, Be and Bt are required to meet normally, the closure of $R(v,e,t)$ is a trihedron with a 2-gonal face in each of the 2-spheres Sv, Se and St.

Let e be an edge of T, and let v_1 and v_2 be the two vertices of e (Fig. 6.47(a)). If e is incident with m triangles $t_1, \ldots, t_m$ of T, then the m triangle spheres $St_1, \ldots, St_m$ of Σ_T meet Se as in Fig. 6.47(b). The closure of $Be - (Bt_1 \cup \cdots \cup Bt_m)$ is another closed 3-disk $\widehat{Be}$ as the one in Fig. 6.47(c). Cutting $\widehat{Be}$ along Sv_1 and Sv_2 (Fig. 6.47(d)) we get a configuration similar to the one in Fig. 6.47(e). Denote by $R(e)$ the region of Σ_T contained in $\widehat{Be} - (Bv_1 \cup Bv_2)$. Its closure is a $2m$-gonal prism, with $2m$-gonal faces in Sv_1 and Sv_2, a 4-gonal face in St_j which is also a face of $R(e, t_j)$ for each $j = 1, \ldots, m$, and another 4-gonal face in $cl(R(\sigma))$ for each tetrahedron $\sigma \in T$ incident with e (Fig. 6.47(e)).

Let $R(v_1, e)$ and $R(v_2, e)$ denote the regions of Σ_T contained in $\widehat{Be} \cap Bv_1$ and $\widehat{Bv} \cap Bv_2$ respectively. For $i = 1, 2$, the closure $R(v_i, e)$ is a "bitten lens", with two $2m$-gonal faces, one of them in Sv_i and the other in Se, and m 2-gonal faces, each of them contained in one of the triangle spheres $St_1, \ldots, St_m$ (Fig. 6.47(e)).

If v is a vertex of T, then $R(v)$ denotes the region of Σ_T containing v. The region $R(v)$ is the open 3-ball $int(Bv)$ with the closure of the regions of the form $R(v, \cdot)$ and $R(v, \cdot, \cdot)$ removed. Figure 6.48(c) shows how the rest of 2-spheres of Σ_T meet Sv when the triangulation T around v is like in Fig. 6.48(a) (see also Fig. 6.41). In this model, the region $R(v)$ is like that in Fig. 6.48(d). The boundary of $R(v)$ has the following shape: if e is an edge of T incident with v and e is incident with m triangles of T, then $R(v)$ has a $2m$-gonal face (which is a face of $R(v, e)$) contained in the edge sphere Se. If t is a triangle of T incident with v, $R(v)$ has a 4-gonal face (which is also a face of $R(v, t)$) in the triangle sphere St. Finally, if σ is a tetrahedron of T incident with v, then $R(v)$ has a 6-gonal face in Sv which is also a face of $R(\sigma)$.

6.4.3 Σ_T *fills* M

The Dehn surface $\Sigma_T = \bigcup_{e \in T^2} Se$ defined above is a filling collection of spheres in M. Moreover, Σ_T is regular and transverse to the triangles of T. By Theorem 2.18, Σ_T proves of Montesinos' Theorem (Theorem 2.11).

It is important to remark that the presented construction, instead of those in [49] or [67], is independent of the orientability of M. Additionally, this construction has the following advantage:

Proposition 6.41. *Let S be a surface and let $f : S \to M$ be a transverse immersion. Let T be a triangulation of M making f simplicial. Then, $f(S) \cup \Sigma_T$ is a regular filling Dehn surface of M.*

Proof. Consider the Dehn surface $\Sigma = f(S)$. Since f is simplicial with respect to T, Σ is contained in the 2-skeleton T^2 of T. The simplices of T^2 meet transversely the spheres of Σ_T, hence Σ meets transversely the 2-spheres of Σ_T, and therefore $\Sigma \cup \Sigma_T$ is a Dehn surface.

Let R be a region of Σ_T. If R does not meet Σ, it is also a region of $\Sigma \cup \Sigma_T$, so let assume that R meets Σ.

If R is a region of the form $R(t)$, $R(\cdot, t)$ or $R(\cdot, \cdot, t)$ for some triangle t of T, then Σ meets R if and only if t is contained in Σ. In this case, Σ divides R into two regions of $\Sigma \cup \Sigma_T$, symmetric with respect to t, whose closures are 6-gonal prisms (if $R = R(t)$), or 3-gonal prisms (if $R = R(\cdot, t)$) or tetrahedra (if $R = R(\cdot, \cdot, t)$) (see Figs. 6.43 and 6.46).

If R is of the form $R(e)$ or $R(\cdot, e)$ for some edge e of T, as in paragraph above, $R \cap \Sigma \neq \emptyset$ if and only if $e \subset \Sigma$. In this case, the edge e is made of either simple or double points of Σ, and then, two (if $e \not\subset \mathrm{Sing}(\Sigma)$) or four (if $e \subset \mathrm{Sing}(\Sigma)$) of the triangles of T incident with e are contained in Σ. Then, by the construction of Σ_T, the region R is divided by Σ into two or four open 3-disks depending on whether $e \not\subset \mathrm{Sing}(\Sigma)$ or $e \subset \mathrm{Sing}(\Sigma)$ (see Figs. 6.42 and 6.47).

If $R = R(v)$ for some vertex v of T, then Σ divides R into two, four or eight open 3-disks depending on whether v is a simple, double or triple point of Σ respectively.

By construction, Σ may meet a face Δ of Σ_T, in this case Σ divides Δ into two or four open 2-disks, and each double curve of $\Sigma \cup \Sigma_T$ contains a triple point.

Therefore, $\Sigma \cup \Sigma_T$ is a filling Dehn sphere of M. By construction, it is straightforward to to show $\Sigma \cup \Sigma_T$ is regular. $\square$

Note that in this proposition, the immersion f can be *any* transverse immersion, filling or not.

Let $f, g : S \to M$ be two transverse immersions related by a pushing disk (D, B), and consider a triangulation T of M making f and g simplicial. Let ϕ_T be a parametrization of Σ_T, and consider the parametrizations $f_T = f \cup \phi_T$ and $g_T = g \cup \phi_T$ of $f(S) \cup \Sigma_T$ and $g(S) \cup \Sigma_T$, respectively. The pushing disk (D, B) also relates the filling immersions f_T and g_T, and since f and g are simplicial with respect to T, the triangulation T induces a triangulation T_B of the pushing ball B.

Key Lemma 2. *If T_B collapses over $g(D)$, f_T is f-homotopic to g_T by a sequence of f-moves and ambient isotopies supported by D.*

The detailed proof of this lemma is long and tedious, so it is postponed until Appendix A.

6.4.4 Inflating filling immersions

Although most of the results of this section are stated in terms of Dehn surfaces, it is possible to write equivalent results for the corresponding parametrizations.

Let Σ be a filling Dehn surface of M, and let $f : S \to M$ be a parametrization of Σ. Consider another Dehn surface Σ' in M such that $\Sigma \cup \Sigma'$ is also a filling Dehn surface of M. Let $\Sigma \# \Sigma'$ be a filling Dehn surface of M obtained from $\Sigma \cup \Sigma'$ performing spiral pipings like one of the following two kinds:

(i) connecting a component of Σ with a component of Σ' around a double point of Σ; or

(ii) connecting two components of Σ' around a triple point of Σ'.

Let $D_\#$ be the union of the piping disks of all these surgeries lying in the domain of Σ.

Definition 6.42. If Σ is f-homotopic to $\Sigma \# \Sigma'$ by a sequence of f-moves and ambient isotopies supported by $D_\#$, then $\Sigma \# \Sigma'$ is an *inflation* of Σ' from Σ. If there exists an inflation of Σ' from Σ, the surface Σ' is said to be *inflatable* from Σ.

Let T be a triangulation of M making f simplicial. Let K be a subset of simplices of T^2 such that $\Sigma \cup \Sigma_K$ fills M. For simplicity of notation, we say that K is *inflatable from* Σ if Σ_K is inflatable from Σ. Accordingly, an inflation of Σ_K from Σ is called an *inflation of K* from Σ. Let K' be another subset of simplices of T^2 such that $\Sigma \cup \Sigma_{K \cup K'}$ fills M.

Definition 6.43. If $\Sigma_{K'-K}$ is inflatable from $\Sigma \cup \Sigma_K$, the complex K' is *inflatable from Σ and K*. An inflation of $\Sigma_{K'-K}$ from $\Sigma \cup \Sigma_K$ is also called an *inflation of K' from Σ and K*.

Lemma 6.44. *If K is inflatable from Σ and K' is inflatable from Σ and K, then $K \cup K'$ is inflatable from Σ.*

Proof. Let $\Sigma \# \Sigma_K$ be an inflation of K from Σ, and let $(\Sigma \cup \Sigma_K) \# \Sigma_{K'}$ be an inflation of K' from Σ and K. All the spiral pipings used to transform $\Sigma \cup \Sigma_K$ into $\Sigma \# \Sigma_K$ are located around triple points of $\Sigma \cup \Sigma_K$, and those used to transform $\Sigma \cup \Sigma_K \cup \Sigma_{K'}$ into $(\Sigma \cup \Sigma_K) \# \Sigma_{K'}$ are located around triple points of $\Sigma \cup \Sigma_K \cup \Sigma_{K'}$ that are not triple points of $\Sigma \cup \Sigma_K$. By Lemma 6.40, the sequence of f-moves transforming $\Sigma \cup \Sigma_K$ into $\Sigma \cup \Sigma_K \cup \Sigma_{K'}$ can be adapted to a sequence of f-moves of $\Sigma \# \Sigma_K$, perhaps replacing some finger moves +2 with piping passing moves. Therefore, the sequence of f-moves transforming Σ into $\Sigma \cup \Sigma_K$ can be extended to a sequence of f-moves deforming Σ into an inflation of $K \cup K'$ from Σ. □

Theorem 6.45. *If T shells Σ, then T^2 is inflatable from Σ. Moreover, there is an inflation $\Sigma \# \Sigma_T$ of T^2 from Σ such that there is only one spiral piping in $\Sigma \# \Sigma_T$ connecting a component of Σ with a component of Σ_T, and the rest of spiral pipings connect different components of Σ_T.*

The remainder of this section is devoted to prove this theorem.

Lemma 6.46 (Inflating triangles). *Let K be a (eventually empty) subcomplex of T^2 such that $\Sigma \cup \Sigma_K$ fills M. Let $t \in T^2 - K$ be a triangle of T, and assume that one of the following conditions holds:*

(a) $t \subset \Sigma$ and there is at least one vertex of T in $K \cup \mathrm{Sing}(\Sigma)$; or
(b) t is exterior to Σ and $\partial t \cap (\Sigma \cup K)$ and $\partial t \cap K$ are connected 1-complexes.

Then, $\Sigma \cup \Sigma_{K \cup \mathrm{cl}(t)}$ fills M and $\mathrm{cl}(t)$ can be inflated from Σ and K. □

The proof of this lemma appears in Appendix 8.3.4.

From now on, when it is said that $\mathrm{cl}(t)$ is inflatable from Σ and K or, equivalently, that t is inflatable from Σ and K, it must be assumed that t meets the hypothesis of the Lemma 6.46 with respect to the filling Dehn surface Σ and the 2-complex K.

Definition 6.47. Let $\mathbf{c} = (t_1, \ldots, t_k)$ be a finite sequence of different triangles of T^2 such that, if

$$K_0 = \emptyset,$$
$$K_i = \mathrm{cl}(t_1) \cup \cdots \cup \mathrm{cl}(t_i), \quad \text{for all } i = 1, \ldots, k,$$

the triangle t_i is inflatable from Σ and K_{i-1} for all $i = 1, \ldots, k$. In this case $\mathbf{c}$ is a *house of cards supported by* Σ. The house of cards $\mathbf{c}$ is *complete* if it contains all the triangles of T^2, and it is *connected* if each triangle t_i has at least one edge in K_{i-1} for $i = 2, \ldots, k$.

Given any sequence $\mathbf{c} = (t_1, \ldots, t_k)$ of triangles of T, define

$$K(\mathbf{c}) = \mathrm{cl}(t_1) \cup \cdots \cup \mathrm{cl}(t_k).$$

The proof of the following proposition is straightforward from Lemma 6.44.

Proposition 6.48. *If $\mathbf{c}$ is a house of cards supported by Σ, then $K(\mathbf{c})$ can be inflated from Σ.* □

Let $c = (t_1, \ldots, t_n)$ be a connected house of cards of T supported by Σ and consider $K = K(c)$. Let $D \subset M$ be a closed 2-disk such that $\text{int}(D) \cap (K \cup \Sigma) = \emptyset$ and D is simplicial with respect to T, and let T_D be the triangulation of D induced by T.

Lemma 6.49. *If* $\partial D \subset K$ *and* s *is any shelling of* T_D*, the concatenation* $c * s$ *is a connected house of cards supported by* Σ.

Proof. Let $s = (\tau_1, \ldots, \tau_m)$ be a shelling of T_D. The proof is by induction on m.

If $m = 1$, then D is the closure of a triangle τ_1 of T with $\partial t \subset K$. By Lemma 6.46, $\text{cl}(\tau_1)$ is inflatable from Σ and K, and this implies that $c * s = (t_1, \ldots, t_n, \tau_1)$ is a connected house of cards supported by Σ.

Assume that the result holds for every disk D' with $m - 1$ triangles, with $m \geq 2$. The triangle τ_1 is free in D, therefore it has one or two edges in $\partial D \subset K$, and the rest of edges in $\text{int}(D)$ which is disjoint from K and Σ. Hence, the intersection of $\partial(\tau_1)$ with $\Sigma \cup K$ coincides with the intersection of $\partial(\tau_1)$ with K, which is a connected 1-complex. By Lemma 6.46, $\text{cl}(\tau_1)$ is inflatable from Σ and K, and $c_1 = (t_1, \ldots, t_n, \tau_1)$ is a connected house of cards supported by Σ. Taking $K_1 = K \cup \text{cl}(\tau_1)$, the set $D_1 = \text{cl}(D - \text{cl}(\tau_1))$ is a closed 2-disk with $\partial D_1 \subset K_1$ and $m - 1$ triangles. Since $s_1 = (\tau_2, \ldots, \tau_m)$ is a shelling of D_1, by the induction hypothesis $c_1 * s_1 = c * s$ is a connected house of cards supported by Σ. $\square$

Assume now that $\partial D \cap K$ is a closed arc γ, and let δ be the complementary arc $\text{cl}(\partial D - \gamma)$ of γ in ∂D.

Lemma 6.50. *Let* s *be a connected shelling of* T_D *going from* γ *to* δ*. If* $\delta \subset \Sigma$ *or* $\text{int}(\delta) \cap \Sigma = \emptyset$*, then* $c * s$ *is a connected house of cards supported by* Σ.

Proof. Let $s = (\tau_1, \ldots, \tau_m)$ be a connected shelling of D going from γ to δ. If $m = 1$, the result follows from Lemma 6.46. Thus, assume that $m > 1$. Set $K_0 = K$ and for $i = 1, \ldots, m$ consider

$$K_i = K \cup \text{cl}(\tau_1 \cup \cdots \cup \tau_i) \text{ and}$$

$$D_i = \text{cl}(\tau_i \cup \cdots \cup \tau_m).$$

Since s is a shelling of T_D, D_i is a 2-disk whose boundary is the union of the two arcs $\gamma_i = \partial D_i \cap K_{i-1}$ and $\delta_i = \partial D_i \cap \delta$, for each $i = 1, \ldots, m$. Since s goes from γ to δ, the triangle τ_1 has at least one edge in $\gamma \subset K = K_0$. Since s is a connected shelling, the triangle τ_i shares an edge with one of the previous triangles $\tau_1, \ldots, \tau_{i-1}$, for all $i > 1$. Hence, $\partial \tau_i$ always has at least one edge in K_{i-1}, for $i = 1, \ldots, m$.

For the last triangle τ_m it is clear that $\partial \tau_m \cap K_{m-1}$ and $\partial \tau_m \cap (\Sigma \cup K_{m-1})$ are connected 1-complexes, therefore τ_m is inflatable from Σ and K_{i-1}

Fix $i \in \{1, \ldots, m-1\}$. By the previous arguments, τ_i has one edge a_i in K_{i-1}, and since $i < m$, it has one edge b_i in $\text{int}(D_i)$. Let c_i be the remaining edge of τ_i, and let A_i be the vertex of τ_i incident with b_i and c_i. If $c_i \subset K_{i-1}$, then

$$\partial \tau_i \cap K_{i-1} = \partial \tau_i \cap (\Sigma \cup K_{i-1}) = \text{cl}(a_i \cup c_i).$$

If $c_i \subset \text{int}(D_i)$, then $A_i \in \text{int}(D_i)$. Otherwise, τ_i is not free in D_i. In this case

$$\partial \tau_i \cap K_{i-1} = \partial \tau_i \cap (\Sigma \cup K_{i-1}) = \text{cl}(a_i).$$

Assume that $c_i \subset \delta_i$. If $A_i \in K_{i-1}$, the arc δ_i coincides with $\text{cl}(c_i)$, in contradiction with the assumption $i < m$. Hence, in this case

$$\partial \tau_i \cap K_{i-1} = \text{cl}(a_i), \qquad \text{and} \qquad \partial \tau_i \cap (\Sigma \cup K_{i-1}) = \text{cl}(a_i \cup c_i).$$

In all these cases the triangle τ_i is inflatable from Σ and K_{i-1}. Therefore, $\mathbf{c} * \mathbf{s}$ is a connected house of cards supported by Σ. □

Let $\mathbf{c}$ and $K(\mathbf{c})$ be as above, and let B be a closed 3-disk in M which is simplicial with respect to T and such that $\partial B \subset K$ and $\text{int}(B) \cap (K \cup \Sigma) = \emptyset$. Consider a cellular decomposition Γ of B such that Γ is also simplicial with respect to T, and denote the 2-skeleton of Γ by Γ^2.

Proposition 6.51. *If Γ is shellable, the triangles of T contained in $\Gamma^2 - \partial B$ can be ordered in a sequence $\mathbf{c}(B)$ such that the concatenation*

$$\mathbf{c} * \mathbf{c}(B)$$

is a connected house of cards supported by Σ.

Proof. Let $(C_1, \ldots, C_k)$ be a shelling of B.

Take $K_0 = K$, and for all $i = 1, \ldots, k-1$ define

$$B_i = \text{cl}(C_i \cup \cdots \cup C_k),$$
$$D_i = \text{cl}(\partial C_i \cap \text{int}(B_i)),$$
$$K_i = K \cup D_1 \cup \cdots \cup D_i.$$

Since $C_1, \ldots, C_k$ is a shelling of Γ, the closure of each 3-cell C_i is free in B_i for all $i = 1, \ldots, k-1$. This implies that $D_1, \ldots, D_k$ are closed 2-disks.

For every $i = 1, \ldots, k-1$, $\partial D_i \subset K_{i-1}$. Therefore, if $\mathrm{sh}(D_i)$ is a shelling of the triangulation of D_i induced by T. by applying Lemma 6.49 inductively we conclude that

$$\mathbf{c} * \mathrm{sh}(D_1) * \cdots * \mathrm{sh}(D_{k-1})$$

is a connected house of cards supported by Σ. It is also clear that

$$D_1 \cup \cdots \cup D_{k-1}$$

agrees with $\mathrm{cl}(\Gamma^2 - \partial B)$, and therefore

$$\mathbf{c}(B) = \mathrm{sh}(D_1) * \cdots * \mathrm{sh}(D_{k-1})$$

satisfies the required conditions. $\qquad\square$

Note that in the previous result, the cell decomposition Γ could coincide with the restriction to B of the triangulation T.

Proof of Theorem 6.45. Let t_1 denote any triangle of T contained in Σ with at least one vertex in $\mathrm{Sing}(\Sigma)$. By Lemma 6.46, t_1 is inflatable from Σ.

Let $t_2, \ldots, t_m$ be the rest of triangles of T contained in Σ. The Dehn surface Σ is filling, so it is a connected subset of M. It is possible order $t_2, \ldots, t_m$ such that each triangle t_i has at least one edge in $\mathrm{cl}(t_1) \cup \cdots \cup \mathrm{cl}(t_{i-1})$, for all $i = 1, \ldots, m$. By Lemma 6.46, $\mathbf{c}(\Sigma) = (t_1, \ldots, t_m)$ is a connected house of cards supported by Σ.

Let R be any region of Σ and assume for simplicity that R is regular.

Since T shells f, considering the cellular decomposition of R induced by T, by Proposition 6.51 there exists a sequence $\mathbf{c}(R)$ of triangles of T containing all the triangles of $T^2 \cap R$ such that

$$\mathbf{c}(\Sigma) * \mathbf{c}(R)$$

is a connected house of cards supported by Σ. If R is not regular and $\hat{q}_R : B^3 \to M$ is a characteristic map of R the sequence $\mathbf{c}(R)$ is constructed in a similar way using the pullback triangulation of B^3 under $\hat{q}_R$.

The construction of $\mathbf{c}(R)$ can be done independently in each region of Σ. If $R_1, \ldots, R_n$ are all the regions of Σ, the concatenation

$$\mathbf{c}(\Sigma) * \mathbf{c}(R_1) * \cdots * \mathbf{c}(R_n)$$

is a complete connected house of cards supported by Σ. By Proposition 6.48, T^2 can be inflated from Σ. By the proof of Lemma 6.46 (see Appendix 8.3.4), the only spiral pipings that must be performed around points of Σ are the first two, and only the first one connects Σ with a component of Σ_T. The rest of spiral pipings are carried out around triple points of Σ_T, so they do not modify Σ. $\qquad\square$

Corollary 6.52. *If $f : S \to M$ is a filling immersion, then f is f-homotopic to a regular filling immersion.*

Proof. By Theorem 6.32, there exists a triangulation T of M that makes f simplicial and shells f. By Theorem 6.45, there exists an immersion f' f-homotopic to f whose image is obtained by applying spiral pipings to $\Sigma \cup \Sigma_T$. By Proposition 6.41, $\Sigma \cup \Sigma_T$ is regular and, by Proposition 2.16, f' is regular.   □

6.5   Filling pairs

Let Σ_1 and Σ_2 be two filling Dehn surfaces of M. Assume, for simplicity, that both are regular.

Proposition 6.53. *If Σ_1 and Σ_2 are disjoint, then M is $\mathbb{S}^3$.*

Proof. If Σ_1 and Σ_2 are disjoint, then Σ_2 is contained in a region R_1 of Σ_1, and Σ_1 is contained in a region R_2 of Σ_2. Therefore, M coincides with the union $R_1 \cup R_2$ of two 3-balls. The result follows from Theorem 6, p. 35 in [56].   □

Definition 6.54. *The filling Dehn surfaces Σ_1 and Σ_2 form a filling pair of M if their union $\Sigma_1 \cup \Sigma_2$ is again a regular filling Dehn surface of M.*

If Σ_1 and Σ_2 are disjoint, there is a region of $\Sigma_1 \cup \Sigma_2$ which is not an open 3-disk (the region $R_1 \cap R_2$ of the proof of Proposition 6.53). Therefore,

Remark 6.55. *If Σ_1 and Σ_2 form a filling pair of M, their intersection is nonempty and transverse.*

Let Σ_1 and Σ_2 form a filling pair of M. Then, Σ_2 induces a cellular decomposition of each region of Σ_1 and vice versa. Moreover, since the union $\Sigma_1 \cup \Sigma_2$ is regular, these induced cellular decompositions are regular too. If R_1 is a region of Σ_1, the Dehn surface Σ_2 is said to *shell* R_1 if the cellular decomposition induced by Σ_2 on $\mathrm{cl}(R_1)$ is shellable. If Σ_2 shells all regions of Σ_1, it is said that Σ_2 *shells* Σ_1.

Definition 6.56. *The Dehn surfaces Σ_1 and Σ_2 are mutually shellable if Σ_1 shells Σ_2 and Σ_2 shells Σ_1.*

Proposition 6.57. *Let Σ_1 and Σ_2 be two regular filling Dehn surfaces of M intersecting transversely. If $f : S_1 \to M$ is a parametrization of Σ_1, then*

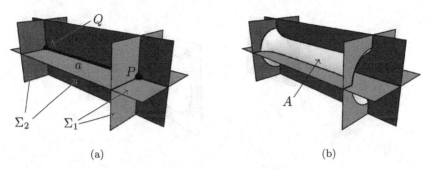

(a) (b)

Fig. 6.49: (a) A bridge between Σ_1 and Σ_3 and, (b) the corresponding type 2 surgery.

f *is f-homotopic to an immersion* $f' : S_1 \to M$ *such that* $\Sigma_1' = f'(S_1)$ *and Σ_2 form a mutually shellable filling pair of M.*

The proof appears in Appendix 8.3.4.

6.6 Simultaneous growings

Assume that Σ_1 and Σ_2 are filling Dehn surfaces of M that form a filling pair, and that there exists two points P and Q near which Σ_1 and Σ_2 intersect as in Fig. 6.49(a). Denote by $\Sigma_1 \# \Sigma_2$ the Dehn surface in M arising from $\Sigma_1 \cup \Sigma_2$ after applying a type 2 surgery between P and Q as in Fig. 6.49(b). If $\Sigma_1 \# \Sigma_2$ is again a filling Dehn surface of M, the edge a of $\Sigma_1 \cup \Sigma_2$ connecting P with Q is said to be a *bridge* between Σ_1 and Σ_2.

If T is a triangulation of M that makes Σ_2 simplicial, it is easy to find a bridge between Σ_T and Σ_2. Moreover, there is always a bridge between the surfaces Σ_1' and Σ_2 of the proof of Proposition 6.57. This implies the following:

Remark 6.58. Up to f-moves in Σ_1 or Σ_2, there is always a bridge between them.

Proposition 6.59. *If Σ_1 and Σ_2 are regular filling Dehn surfaces that form a filling pair, then there exists a bridge between them.*

Proof. Since Σ_1 and Σ_2 form a filling pair, they are not disjoint and their intersection is transverse. Let L be a connected component of $\Sigma_1 \cap \Sigma_2$. Each triple point of $\Sigma_1 \cup \Sigma_2$ belonging to L is either a double point of Σ_1 or a double point of Σ_2. If L contains no double point of Σ_1, it is contained in

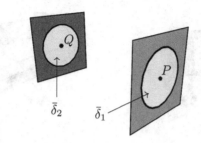

Fig. 6.50

a face of Σ_1 and $\Sigma_1 \cup \Sigma_2$ cannot be filling. Therefore, L contains double points of Σ_1 and, in the same way, it contains double points of Σ_2.

Since L is connected, there exists an edge a of $\Sigma_1 \cup \Sigma_2$ contained in L that connects a double point of Σ_1 with a double point of Σ_2. The configuration of $\Sigma_1 \cup \Sigma_2$ around a is like the one depicted in Fig. 6.49(a). It follows from the regularity of both Σ_1 and Σ_2 that the surface $\Sigma_1 \# \Sigma_2$ obtained from $\Sigma_1 \cup \Sigma_2$ by a type 2 surgery along a is filling. $\square$

Let a be a bridge between Σ_1 and Σ_2 as above. Let $f : S_1 \to M$ and $g : S_2 \to M$ be parametrizations of Σ_1 and Σ_2 respectively, and let $f \# g$ be the parametrization of $\Sigma_1 \# \Sigma_2$ obtained from $f \cup g$ by a type 2 surgery along a. Let T be a triangulation of M that makes f, g and $f \# g$ simplicial, and assume that T shells f. Consider an inflation f' of T from f as in Theorem 6.45, such that f' agrees with f on S_1 except in a small 2-disk $D' \subset S_1$, and let denote $\Sigma_1' = f'(S_1)$. Take D' such that its image under f does not intersect the piping A along the bridge a. In this situation, the piping along a can also be considered as connecting $f'(S_1)$ and Σ_2, and it is possible to consider the immersion $f' \# g : S_1 \# S_2 \to M$ which agrees with f' on $S_1 - \delta_1$ and with g on $S_2 - \delta_2$, where $\delta_1 \subset S_1$ and $\delta_2 \subset S_2$ are the piping disks of the surgery.

Key Lemma 3. *If Σ_2 shells Σ_1, we can pick T and f' such that $f' \# g$ is an inflation of T from $f \# g$.*

Proof. Let $D_1, \ldots, D_{14}$ be the closed 2-disks involved in the surgery as depicted in Fig. 6.51:

- $D_1, \ldots, D_4$ are the four 3-gons into which $\Sigma_1 \cup \Sigma_2$ splits $\bar{\delta}_1 = f(\delta_1)$.
- $D_5, \ldots, D_8$ are the four 4-gons into which $\Sigma_1 \cup \Sigma_2$ splits the piping A.

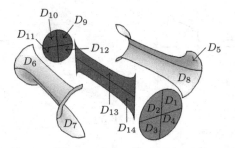

Fig. 6.51: The disks involved in a type 2 surgery.

- $D_9, \ldots, D_{12}$ are the four 3-gons into which $\Sigma_1 \cup \Sigma_2$ splits $\bar{\delta}_2 = f(\delta_2)$.
- D_{13}, D_{14} are the two 4-gons into which Σ_1 splits the subset of Σ_2 interior to the piping cylinder.

All these disks are simplicial with respect to T. Put $K_0 = f\left(\mathrm{cl}(S_1 - \delta_1)\right)$ and set

$$K_i = K_{i-1} \cup D_i, \quad \text{for each } i = 1 \ldots, 14.$$

Arguments like those at the first part of the proof of Theorem 6.45 imply that K_0 is inflatable from Σ_1 and from $\Sigma_1 \# \Sigma_2$. Moreover, by subdividing T if necessary in order to guarantee that the restrictions of T to the 2-disks $D_1, \ldots, D_{14}$ have shellings with the required properties, using Lemmas 6.49 and 6.50 and Proposition 6.48, it can be checked that D_i *is inflatable from Σ_1 and K_{i-1}, and from $\Sigma_1 \# \Sigma_2$ and K_{i-1} for each* $i = 1, \ldots, 14$. Therefore, K_{14} is inflatable from Σ_1 and $\Sigma_1 \# \Sigma_2$. This construction provides a connected house of cards $\mathbf{c}_0$ supported by Σ_1 and from $\Sigma_1 \# \Sigma_2$ such that $K(\mathbf{c}_0) = K_{14}$. Denote K_{14} simply by K.

Let $R_1, \ldots, R_m$ be the connected components of $M - K$ that have nonempty intersection with Σ_2. Since Σ_2 shells Σ_1, the cell decomposition Γ_1 that Σ_2 induces in $\mathrm{cl}(R_1)$ is shellable. By Proposition 6.51, there is a sequence $\mathbf{c}_1$ containing all the triangles of T lying in $\Sigma_2 \cap R_1$ such that the concatenation $\mathbf{c}_0 * \mathbf{c}_1$ is a connected house of cards supported by Σ_1. Besides that, since $K(\mathbf{c}_1) \subset \Sigma_2$, the sequence $\mathbf{c}_0 * \mathbf{c}_1$ is also a connected house of cards supported by $\Sigma_1 \# \Sigma_2$. Applying the same argument with $R_2, \ldots, R_m$, we get sequences $\mathbf{c}_2, \ldots, \mathbf{c}_m$ such that $K(\mathbf{c}_j) = \mathrm{cl}(\Sigma_2 \cap R_j)$, for each $j = 2, \ldots, m$, and such that

$$\mathbf{d} = \mathbf{c}_0 * \mathbf{c}_1 * \cdots * \mathbf{c}_m$$

is a connected house of cards supported by Σ_1 and by $\Sigma_1 \# \Sigma_2$.

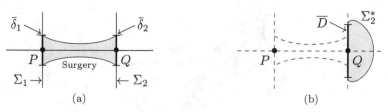

Fig. 6.52

By construction $K(\mathbf{d}) = \Sigma_1 \cup \Sigma_2 \cup A$. Let $R'_1, \ldots, R'_n$ be the connected components of $M - K(\mathbf{d})$. It can be assumed again that T is sufficiently subdivided to guarantee that the cell decomposition induced by T in the closure of R'_i is shellable, for each $i = 1, \ldots, n$. By succesive applications of Proposition 6.51, $\mathbf{d}$ can be extended to a connected house of cards $\mathbf{c}$ supported by Σ_1 and by $\Sigma_1 \# \Sigma_2$ which is complete, that is, such that $K(\mathbf{c}) = T^2$.

Finally, modifying f by successive applications of Lemma 6.46 following the house of cards $\mathbf{c}$, an inflation f' of T from f is obtained. Since the house of cards $\mathbf{c}$ is also supported by $\Sigma_1 \# \Sigma_2$, by the proof of Lemma 6.46, the modifications applied to f can also be seen as modifications of $f \# g$ by f-moves. Therefore, the inflation f' of T from f verifies that $f' \# g$ is also an inflation of T from $f \# g$. □

6.7   Proof of Theorem 5.8

Let Σ_1 and Σ_2 be a pair of null-homotopic filling Dehn spheres of M. Let $f : S_1 \to M$ and $g : S_2 \to M$ be parametrizations of Σ_1 and Σ_2 respectively.

Modifying f if necessary by ambient isotopy, it is possible to assume that Σ_1 and Σ_2 have nonempty intersection and that they meet transversely.

By Propositions 6.57 and 6.59, modifying f with f-moves if necessary, it is possible to suppose that Σ_1 and Σ_2 form a mutually shellable filling pair in M and that there exists a bridge a between them (Figs. 6.53(a) to 6.53(d)). Consider the filling Dehn sphere $\Sigma_1 \# \Sigma_2$ obtained from $\Sigma_1 \cup \Sigma_2$ by a type 2 surgery along a, and let $f \# g$ be the parametrization of $\Sigma_1 \# \Sigma_2$ obtained from $f \cup g$ by the type 2 surgery along a. Consider also the piping disks $\delta_1 \subset S_1$ and $\delta_2 \subset S_2$ and their images $\bar{\delta}_1 = f(\delta_1)$ and $\bar{\delta}_2 = g(\delta_2)$ and the piping A as in Sec. 6.6 (see Fig. 6.49(b)).

Consider a small 2-sphere Σ_2^* standardly embedded in M as in Fig. 6.53(e) (see also Fig. 6.52) and a parametrization $g_* : S_2 \to M$ of Σ_2^*. This 2-sphere shares with Σ_2 a closed disk $\overline{D}$ containing $\bar{\delta}_2$ in its interior. Assume that

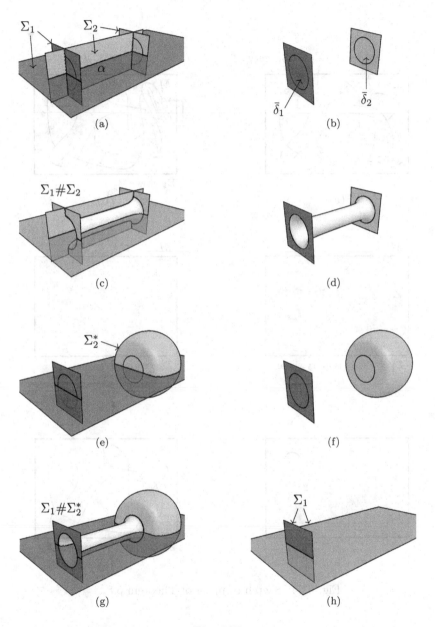

Fig. 6.53

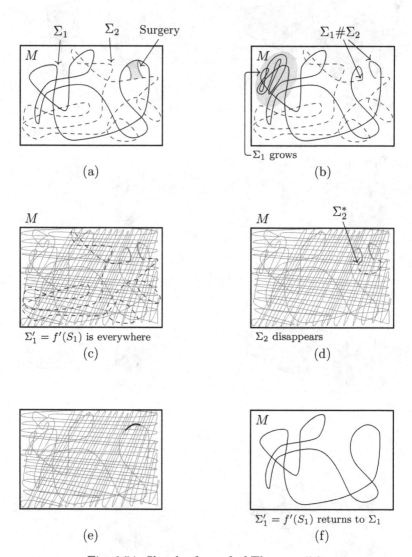

Fig. 6.54: Sketch of proof of Theorem 5.8.

the immersions g and g_* agree on the 2-disk of $g^{-1}(\overline{D})$ that contains δ_2. By Key Lemma 1, g can be deformed into g_* by a finite sequence of transverse pushing disks fixing $\bar{\delta}_2$. Let $(D_1, B_1), \ldots, (D_k, B_k)$ be this sequence of pushing disks, and let $g = g_0, g_1, \ldots, g_k = g_* : S_2 \to M$ be the sequence of transverse immersions such that g_i is obtained from g_{i-1} by the pushing disk (D_i, B_i), for all $i = 1, \ldots, k$.

Slightly modifying Σ_1 and/or the piping A between Σ_1 and Σ_2 by ambient isotopy if necessary, all these pushing disks $(D_1, B_1), \ldots, (D_k, B_k)$ can be assumed to be transverse to Σ_1 and $\Sigma_1 \# \Sigma_2$. Since all the pushing disks fix δ_2, they can be regarded as acting on the immersion $f \# g$ instead of on g. Hence, in the sequence of transverse immersions

$$f \# g = f \# g_0, f \# g_1, \ldots, f \# g_k = f \# g_* : S_1 \# S_2 \to M,$$

the immersion $f \# g_i$ is obtained from $f \# g_{i-1}$ by the pushing disk (D_i, B_i), for all $i = 1, \ldots, k$. Notice that $f \# g_k$ parametrizes $\Sigma_1 \# \Sigma_2^*$, where $\Sigma_1 \# \Sigma_2^*$ is obtained connecting Σ_1 with Σ_2^* with the piping A (Fig. 6.53(g)). A last transverse pushing disk (D_{k+1}, B_{k+1}) transforms $f \# g_k$ (Fig. 6.53(g)) into f (Fig. 6.53(h)). Therefore, *there exists a finite sequence of transverse pushing disks fixing* $\mathrm{cl}(S_1 - \delta_1)$ *that transforms* $f \# g$ *into* f.

Consider a good triangulation T of M with respect to f, g, $g_1, \ldots, g_k$, $f \# g$, $f \# g_1, \ldots, f \# g_k$. Since Σ_1 and Σ_2 form a mutually shellable filling pair, Σ_2 shells Σ_1. Subdividing T if necessary, take an inflation f' of T from f and the immersion $f' \# g$ in the conditions of Key Lemma 3. Consider now the sequence of immersions

$$f' \# g = f' \# g_0, f' \# g_1, \ldots, f' \# g_k = f' \# g_* : S_1 \# S_2 \to M,$$

such that, for each $i = 1, \ldots, k + 1$, the immersion $f' \# g_i$ is obtained from $f' \# g_{i-1}$ by the pushing disk (D_i, B_i). By the election of T, the triangulation T restricted to the pushing ball B_i collapses over $g_i(D_i) = f' \# g_i(D_i)$ for each $i = 1, \ldots, k$. By Key Lemma 2 and Lemma 6.40, this means that $f' \# g_i$ is f-homotopic to $f' \# g_{i-1}$ for each $i = 1, \ldots, k + 1$. The same argument applies to the pushing disk (D_{k+1}, B_{k+1}) that transforms $f' \# g_*$ into f'. Therefore $f' \# g_*$ is f-homotopic to f'. In conclusion (see Fig. 6.54),

(i) $f \# g$ is f-homotopic to $f' \# g$, since $f' \# g$ is an inflation of T from $f \# g$;

(ii) $f' \# g$ is f-homotopic to f', by repeated application of Key Lemma 2 to the pushing disks (D_i, B_i) for $i = 1, \ldots, k + 1$; and

(iii) f' is f-homotopic to f, because f' is an inflation of T from f.

Hence, $f \# g$ is f-homotopic to f. For the same reasons, $f \# g$ is f-homotopic to g, and hence f and g are f-homotopic. $\qquad \square$

Chapter 7

The triple point spectrum

7.1 The Shima's spheres

Consider a diagram $\mathcal{D}$ in $\mathbb{S}^2$ of a pseudo Dehn sphere Σ with p triple points. Each crossing of $\mathcal{D}$ is the intersection of two small arcs contained in one or two curves of the diagram. Let call *crossing arcs* to these arcs. It is well-known that the number of crossings of two transverse closed curves in $\mathbb{S}^2$ is even. We refer to this property as the *even intersection property*, and it implies that the number of crossing arcs contained in a single curve of $\mathcal{D}$ must be even: each self-intersection uses two crossing arcs and by the even intersection property the number of crossings arcs that intersect other curves of the diagram must be even too. Since there is an even number of curves in $\mathcal{D}$, the number of crossing arcs must be a multiple of 4. If the diagram has $3p$ crossings, the number of crossing arcs is $6p$, and therefore p is even. Summarizing:

Lemma 7.1. *Any (pseudo) Dehn sphere has an even number of triple points.*

Hence, a filling Dehn sphere of a 3-manifold has at least two triple points. In [62], Shima classifies the Dehn spheres in $\mathbb{S}^3$ with 2 triple points, introducing six examples of such Dehn spheres. Three of them are filling, and the main Theorem of [62] implies that they are the only three possible ones. These three spheres are the following:

Johansson's sphere. Whose diagram, depicted in Fig. 3.2(a), was introduced in [36] (Example 1.5 of [62]);

Montesinos' sphere. Obtained applying to $\mathbb{S}^3$ the algorithm introduced in [49] to obtain filling Dehn spheres (Example 1.6 of [62]). Figure 7.1(a) shows its diagram;

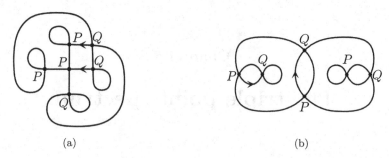

(a)   (b)

Fig. 7.1: Montesinos' and Banchoff's sphere diagrams.

Banchoff's sphere. Obtained applying two type 1 surgeries to the triple ball depicted in Fig. 2.3(b) (Example 1.3 again in [62]). Figure 7.1(b) shows its diagram.

Since every Dehn surface in $\mathbb{S}^3$ is null-homotopic, Johansson's, Montesinos' and Banchoff's spheres are null-homotopic. Theorem 5.22 implies that the diagrams of Figs. 3.2(a), 7.1(a) and 7.1(b) can be related using f-moves. Figures 7.2 and 7.3 show how these diagram are related by f-moves. A class of neighbouring curves has been marked in gray in each diagram of these figures.

Apply a finger move $+1$ along the arc α to the Banchoff's sphere diagram (Fig. 7.2(a)) to obtain Fig. 7.2(b). Modifying this diagram using a saddle move along the arcs β and γ, Fig. 7.2(c) is obtained. The diagram of Fig. 7.2(c) only has two double curves and it is realizable. By Corollary 4.9 it is a filling diagram and this saddle move is an f-move. Modifying now Fig. 7.2(c) with a saddle move along the arcs δ and ε, Fig. 7.2(d) is obtained. This diagram is isotopic to the one in Fig. 7.2(e). Finally, apply to this last diagram a finger move -1 to get Fig. 7.2(f). Figure 7.2(f) is the diagram of the Johansson's sphere (see Fig. 7.3(a)), and therefore this diagram is filling. Additionally, since Fig. 7.2(e) is obtained from Fig. 7.2(f) by a finger move $+1$ along the arc λ depicted in this last figure, the diagram of Fig. 7.2(e) is also filling. Then, all the diagrams in Fig. 7.2 are filling, and therefore all these moves are f-moves. On the other hand, starting with the diagram of the Johansson's sphere as depicted in Fig. 7.3(a), applying a saddle move along the arcs μ and ρ, Fig. 7.3(b) is obtained. Modifying this last diagram by isotopy, we get the diagram of Fig. 7.3(c), which is isotopic to the diagram of the Montesinos' sphere depicted in Fig. 7.3(d).

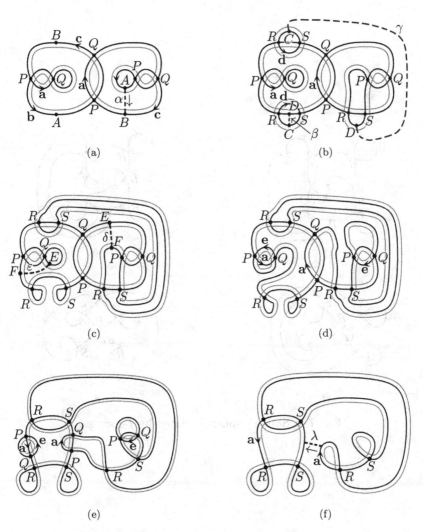

Fig. 7.2: Shima's spheres related by f-moves.

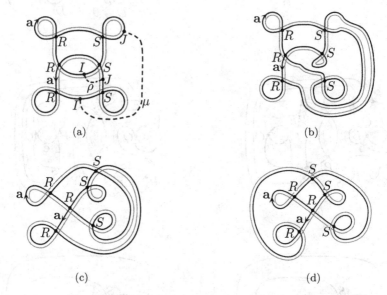

Fig. 7.3: Shima's spheres related by **f**-moves.

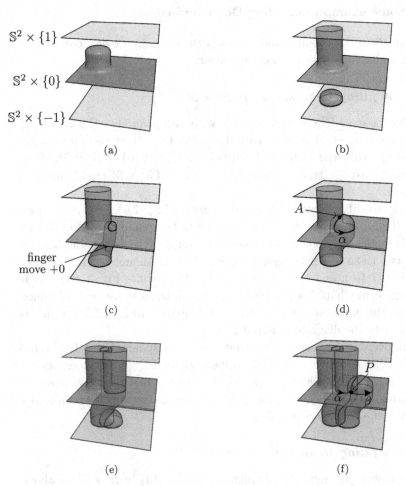

$\mathbb{S}^2 \times \{1\}$

$\mathbb{S}^2 \times \{0\}$

$\mathbb{S}^2 \times \{-1\}$

(a)

(b)

finger
move $+0$

(c)

A

α

(d)

(e)

P

α β

(f)

Fig. 7.4: Deformation of an embedded sphere in $\mathbb{S}^2 \times \mathbb{S}^1$.

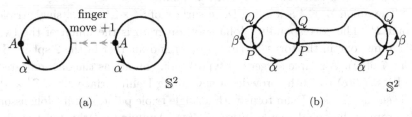

finger
move $+1$

A

α

A

α

$\mathbb{S}^2$

(a)

Q

β

P

Q

P

α

Q

β

P

α

$\mathbb{S}^2$

(b)

Fig. 7.5: A diagram on $\mathbb{S}^2$ representing $\mathbb{S}^2 \times \mathbb{S}^1$.

7.2   Some examples of filling Dehn surfaces

Through the section, points and curves in the diagram are denoted as their images under the corresponding immersions.

7.2.1   *A filling Dehn sphere in* $\mathbb{S}^2 \times \mathbb{S}^1$

Consider the space $\mathbb{S}^2 \times \mathbb{S}^1$ as $\mathbb{S}^2 \times [-1, 1]$ with each point $(x, -1) \in \mathbb{S}^2 \times \{-1\}$ of the *lower sheet* identified with the point $(x, +1)$ of the *upper sheet* $\mathbb{S}^2 \times \{+1\}$. Take the embedded 2-sphere $\Sigma = \mathbb{S}^2 \times \{0\}$ in $\mathbb{S}^2 \times \mathbb{S}^1$. Now, it is possible to construct a filling Dehn sphere of $\mathbb{S}^2 \times \mathbb{S}^1$ modifying Σ as follows.

First, modify Σ by an ambient isotopy (Fig. 7.4). Starting from Fig. 7.4(b), a finger move $+0$ leads to the Dehn sphere Σ_1 of Fig. 7.4(c), which is ambient isotopic to the one of Fig. 7.4(d). The Johansson diagram of this last Dehn sphere is depicted in Fig. 7.5(a). An ambient isotopy take us to Fig. 7.4(e), and then a finger move $+1$ provides Fig. 7.4(f). As it is indicated in Chap. 5.4 (Fig. 5.12(a)), the Johansson diagram obtained after the this finger move $+1$ is the one depicted in Fig. 7.5(b), which is equivalent to the diagram of Fig. 3.2(f).

Moreover, it can be checked that the Dehn sphere Σ_2 of Fig. 7.4(f) fills $\mathbb{S}^2 \times \mathbb{S}^1$. However, the Dehn sphere Σ_1 of Fig. 7.4(e) does not fill $\mathbb{S}^2 \times \mathbb{S}^1$ because it has no triple points. This means that the finger move -1 transforming Σ_2 into Σ_1 *is not* an f-move. This is an explicit example of a finger move -1 which is not an f-move.

7.2.2   *A filling Dehn torus in* $\mathbb{S}^2 \times \mathbb{S}^1$

Consider also the embedded 2-sphere $\Sigma = \mathbb{S}^2 \times \{0\}$ in $\mathbb{S}^2 \times \mathbb{S}^1$ as above. Take a lemniscate in Σ as in Fig. 7.6(a), and consider its product with the vertical component $\mathbb{S}^1$ (Fig. 7.6(b)) to obtain a Dehn torus Σ' in $\mathbb{S}^2 \times \mathbb{S}^1$ with a single double curve and no triple points.

The union $\Sigma \cup \Sigma'$ is a filling Dehn surface of $\mathbb{S}^2 \times \mathbb{S}^1$ with a single triple point P. The corresponding Johansson diagram is the union of the two diagrams, one in the torus $\mathbb{T}^2$ (Fig. 7.6(c)) and another in the 2-sphere $\mathbb{S}^2$ (Fig. 7.6(d)). A simple surgery of type 1 along an arc as the one indicated in Figs. 7.6(c) to 7.6(h) provides a new filling Dehn surface Σ^* of $\mathbb{S}^2 \times \mathbb{S}^1$. The surface Σ^* is a Dehn torus with a single triple point, and its Johansson diagram is the one depicted in Fig. 7.7(a). Applying a Dehn twist to the abstract 2-torus $\mathbb{T}^2$, we obtain the equivalent diagram of Fig. 7.7(b).

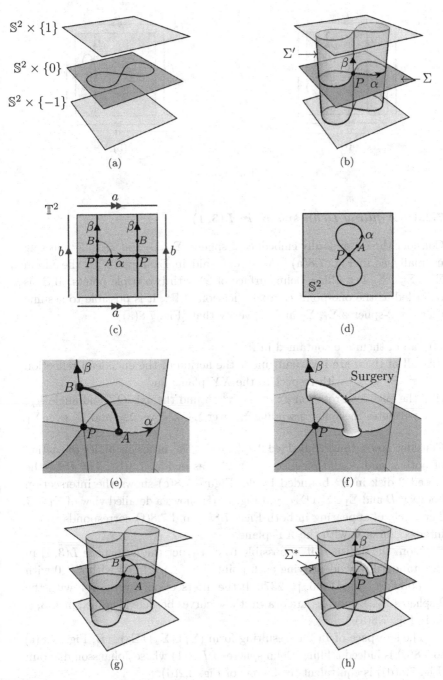

Fig. 7.6: A filling Dehn torus in $\mathbb{S}^2 \times \mathbb{S}^1$.

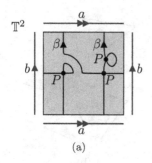
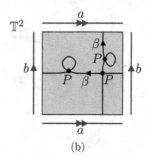

(a)   (b)

Fig. 7.7

7.2.3   A filling Dehn sphere in $L(3,1)$

Consider three standardly embedded 2-spheres Σ_1, Σ_2 and Σ_3 in $\mathbb{S}^3$ meeting normally as in Fig. 7.8(a). As it was said in Example 2.10, the union $\Sigma_1 \cup \Sigma_2 \cup \Sigma_3$ is a filling Dehn surface of $\mathbb{S}^3$ with two triple points. If $\mathbb{S}^3$ is regarded as the one-point compactification of $\mathbb{R}^3$, it is possible to assume that the 2-spheres Σ_1, Σ_2 and Σ_3 verify that (Fig. 7.8(a)):

(i) all of them are contained in $\mathbb{R}^3$;

(ii) all of them are invariant under the action of the euclidean reflection $\pi : \mathbb{R}^3 \to \mathbb{R}^3$ with respect to the XY plane; and

(iii) the euclidean rotation $\rho : \mathbb{R}^3 \to \mathbb{R}^3$ around the axis OZ and angle $2\pi/3$ permutes the spheres sending Σ_1 over Σ_2, Σ_2 over Σ_3, and Σ_3 over Σ_1.

Consider now a third embedded 2-sphere $\Phi \subset \mathbb{R}^3$ invariant under the action of both π and ρ, and meeting $\Sigma_1, \Sigma_2, \Sigma_3$ as in Fig. 7.8(b). Let B be the closed 3-disk in $\mathbb{R}^3$ bounded by Φ. Figure 7.8(c) shows the intersection between B and $\Sigma_1 \cup \Sigma_2 \cup \Sigma_3$, and Fig. 7.8(f) shows a detailed view of $\Sigma_1 \cap B$. The circle C appearing in both Figs. 7.8(c) and 7.8(f) corresponds to the intersection of Φ with the XY plane.

From the 3-disk B it is possible to construct the lens space $L(3,1)$ in the standard way, identifying each point $A = (x,y,z) \in \Phi$ with $z \leq 0$ with the point $(\pi \circ \rho)(A)$ [56, p. 237]. If the intersection curves of Φ with the 2-spheres Σ_1, Σ_2 and Σ_3 are drawn, these curves become identified in $L(3,1)$ as in Fig. 7.8(a).

The subspace of $L(3,1)$ resulting form $(\Sigma_1 \cup \Sigma_2 \cup \Sigma_3) \cap B$, Figs. 7.8(c) to 7.8(d) is indeed a filling Dehn sphere of $L(3,1)$ whose Johansson diagram (Fig. 7.8(d)) is equivalent to the one of Fig. 3.2(d).

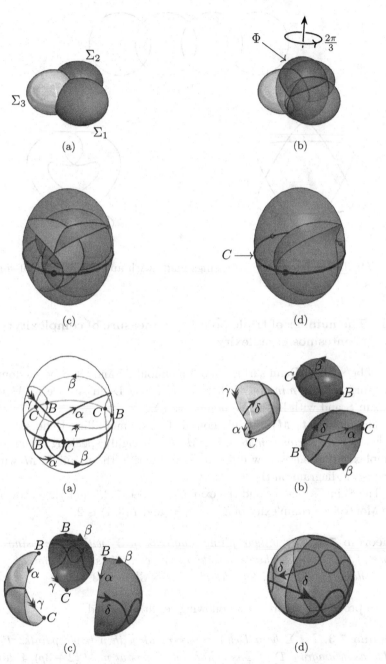

Fig. 7.8: Building a filling Dehn sphere in $L(3,1)$.

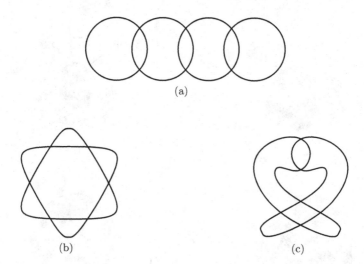

(a)

(b)   (c)

Fig. 7.9: Diagrams with six crossings made with an even number of simple curves.

7.3   The number of triple points as a measure of complexity: Montesinos complexity

Let Σ be a filling Dehn surface in a 3-manifold M and let S be its domain. The surface Σ is *minimal* if there is no filling Dehn surface in M with domain S and with less triple points than Σ.

The number $t_0(M)$ of triple points of a minimal filling Dehn sphere of M is the *Montesinos complexity of M* [43]. Roughly speaking, Montesinos complexity measures how difficult is to describe the manifold M with a Johansson diagram on the 2-sphere.

The Shima's spheres and the examples of Sec. 7.2.1 and 7.2.3 show that the Montesinos complexity of $\mathbb{S}^3$, $\mathbb{S}^2 \times \mathbb{S}^1$ and $L(3,1)$ is 2.

Theorem 7.2. *The unique filling diagrams in $\mathbb{S}^2$ with six crossings are those of the Shima's spheres and the examples of Secs. 7.2.1 and 7.2.3. In particular, if $t_0(M) = 2$, then M is $\mathbb{S}^3$, $\mathbb{S}^2 \times \mathbb{S}^1$ or $L(3,1)$.*

To prove this theorem the following result is needed.

Lemma 7.3. *Let Σ be a Dehn sphere in M with p triple points. If the Johannson diagram $\mathcal{D}$ of Σ is connected, Σ has at most $(2+3p)/4$ double curves.*

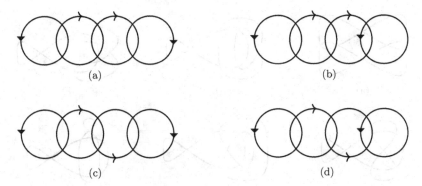

Fig. 7.10: Diagrams supported by the graph of Fig. 7.9(a).

Proof. Although $\mathcal{D}$ has $3p$ crossings, by the even intersection property it has at most $3p/2$ distinct pairs of curves with nonempty intersection. Since $\mathcal{D}$ is connected, it has at most $1 + 3p/2$ different curves. Therefore, Σ has at most $(1 + 3p/2)/2 = (2 + 3p)/4$ double curves. □

Proof of Theorem 7.2. The proof is elementary but quite long, so the details are left to the reader. It consists in the following steps:

(i) make a list of all the diagrams in $\mathbb{S}^2$ with 6 crossings;

(ii) check which of these diagrams are realizable; and

(iii) among the realizable ones, check which diagrams are filling.

We sketch all these steps in the following paragraphs.

Let Σ be a filling Dehn sphere with 2 triple points, and let $\mathcal{D}$ be its Johansson diagram. By Lemma 7.3, Σ has at most two double curves.

If Σ has two double curves, $|\mathcal{D}|$ is, up to isotopy, the "audi" graph of Fig. 7.9(a). This graph supports exactly four diagrams, which can be seen in Fig. 7.10. According to Theorem 3.15, among these diagrams, only the ones in Figures 7.10(b) and 7.10(c) are realizable, and only Fig. 7.10(b) is filling (see Example 3.26). Therefore, in this case Σ is the filling Dehn sphere of $\mathbb{S}^2 \times \mathbb{S}^1$ given in Sec. 7.2.1. It must be noted that the diagrams of Figs. 7.10(a) and 7.10(d) are realizable in a non-orientable 3-manifold, and that Fig. 7.10(a) is filling.

If Σ has one double curve, the Johansson diagram of Σ has two curves α and $\tau\alpha$. Both curves α and $\tau\alpha$ are simple, or both are non-simple.

If α and $\tau\alpha$ are simple, $|\mathcal{D}|$ is, up to isotopy, the graph of Fig. 7.9(b) or Fig. 7.9(c). No diagram $\mathcal{D}$ in $\mathbb{S}^2$ has Fig. 7.9(c) as its graph $|\mathcal{D}|$. On the other hand, there is only one diagram $\mathcal{D}$ such that $|\mathcal{D}|$ is the graph of

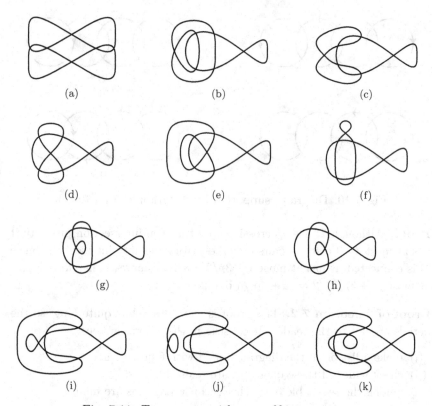

Fig. 7.11: Two curves with one self-intersection.

Fig. 7.9(b): the diagram of the filling Dehn sphere of $L(3,1)$ introduced in Sec. 7.2.3 (see Figs. 7.8(d) and 3.2(d)).

Assume that α and $\tau\alpha$ are non-simple. If both α and $\tau\alpha$ have one crossing, $|\alpha| \cap |\tau\alpha|$ contains four points and $|\mathcal{D}|$ must be one of the graphs depicted in Fig. 7.11. None of them supports any diagram.

If both α and $\tau\alpha$ have two crossing, $|\alpha| \cap |\tau\alpha|$ contains two points and $|\mathcal{D}|$ must be one of the graphs depicted in Fig. 7.12. From these graphs, only the diagrams in Fig. 7.13 can be obtained. Each of these three diagrams can be embedded in $\mathbb{S}^2$ in many ways. For example, Fig. 7.14 shows the other possible embeddings of the diagram of Fig. 7.13(a) in $\mathbb{S}^2$. For each of these embeddings, the realizability condition of Theorem 3.15 must be checked. The unique realizable diagrams coming from the graphs of Fig. 7.13 are exactly the diagrams of the Shima's spheres. By Corollary 4.9, these diagrams are filling. □

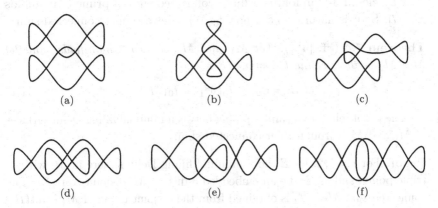

Fig. 7.12: Two curves with two self-intersections.

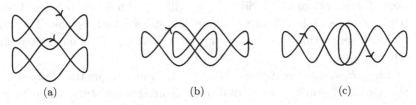

Fig. 7.13: Diagrams with two curves with self-intersections.

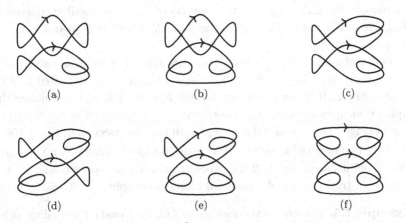

Fig. 7.14: Other embeddings in $\mathbb{S}^2$ of the diagram of Johansson's sphere.

Any closed 3-manifold is a finite connected sum of prime 3-manifolds [27, 47]. So it is natural to ask how $t_0(M)$ behaves under connected sum.

Theorem 7.4 (cf. [43]). *Let M_1 and M_2 be two 3-manifolds, and let $M_1 \# M_2$ be their connected sum. Then*

$$t_0(M_1 \# M_2) \leq t_0(M_1) + t_0(M_2) + 2. \tag{7.1}$$

The proof relays in a careful piping between minimal filling Dehn surfaces of M_1 and M_2 through their connected sum.

Proof. For $i = 1, 2$, let Σ_i be a minimal filling Dehn sphere in M_i with p_i triple points and let B_i be an embedded 3-disk in M_i disjoint from Σ_i. The connected sum $M_1 \# M_2$ is obtained from the disjoint union of $M_1 - \text{int}(B_1)$ and $M_2 - \text{int}(B_2)$ when the boundaries of B_1 and B_2 become identified by a homeomorphism.

Recall that a face of a Dehn surface Σ_i is separating if it is incident with two different regions of Σ_i. Since Σ_i is a filling Dehn sphere it has at least 2 triple points and, by Theorem 3.25, it has at least four regions. Therefore Σ_i has a separating face. Moreover, each region of Σ_i is incident with a separating face.

Since B_i and Σ_i are disjoint, for $i = 1, 2$, Σ_1 and Σ_2 are transformed after the connected sum into a pair of disjoint Dehn spheres, still denoted Σ_1 and Σ_2, of $M_1 \# M_2$, and the connected component R of $(M_1 \# M_2) - (\Sigma_1 \cup \Sigma_2)$ lying between them is homeomorphic to $\mathbb{S}^2 \times I$, where I is an open interval. Take a simple type 0 arc a in $M_1 \# M_2$ with respect to $\Sigma_1 \cup \Sigma_2$ that connects a separating face Δ_1 of Σ_1 with a face Δ_2 of Σ_2. It is possible to assume that a is *unknotted* in R, that is, the pair $(R, a \cap R)$ is homeomorphic to the pair $(\mathbb{S}^2 \times I, \{*\} \times I)$.

Performing on $\Sigma_1 \cup \Sigma_2$ a type 0 surgery along a, a Dehn sphere $\Sigma_1 \# \Sigma_2$ is obtained (Fig. 7.15(b)). Since Δ_1 is separating and a is unknotted $\Sigma_1 \# \Sigma_2$ is quasi-filling. It is not filling because the face of $\Sigma_1 \# \Sigma_2$ that contains the piping is an open annulus. This obstruction can be removed by throwing two fingers (Fig. 7.15(c)) along the piping, until they intersect as in Fig. 7.15(d), creating two new triple points. The resulting Dehn sphere $\widetilde{\Sigma_1 \# \Sigma_2}$ is now filling, and it has $p_1 + p_2 + 2$ triple points, which is an upper bound for the number of triple points of a minimal filling Dehn sphere in $M_1 \# M_2$. $\quad\square$

Example 7.5. Let M_1 be the lens space $L(3, 1)$. Consider the filling Dehn sphere Σ_1 of M_1 described in Sec. 7.2.3 and an embedded 3-disk $B_1 \subset M_1$ disjoint from Σ_1. Since Σ_1 has 2 triple points, it is minimal and the

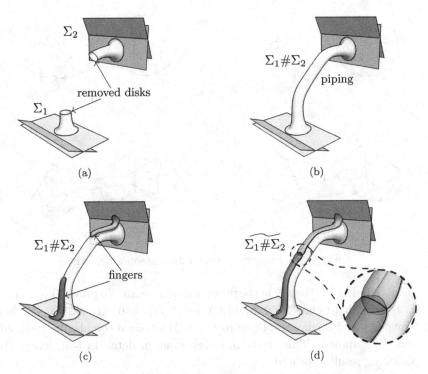

Fig. 7.15: Surgery between disjoint Dehn surfaces.

Montesinos complexity of $L(3,1)$ is 2. Take an exact copy (M_2, Σ_2, B_2) of (M_1, Σ_1, B_1) and construct $L(3,1)\#L(3,1)$ by means of the identity map id $: \partial B_1 \to \partial B_2$. Applying the theorem above,

$$t_0\big(L(3,1)\#L(3,1)\big) \leq 6.$$

Figure 7.16 shows the modifications made on two copies of the Johansson diagram of Σ in order to obtain the Johansson diagram of the filling Dehn sphere $\widetilde{\Sigma_1\#\Sigma_2}$ of $L(3,1)\#L(3,1)$ costructed from Σ_1 and Σ_2 as in the proof of Theorem 7.4. There are two copies of the diagram of Fig. 3.2(d) on two copies of $\mathbb{S}^2$. With the previous assumptions, Σ_1 and Σ_2 are two specular copies of each other in $L(3,1)\#L(3,1)$. If the disks removed from Σ_1 and Σ_2 during the surgery were also identical, two specular copies of the same Johansson diagram are pasted to obtain the Johansson diagram of $\widetilde{\Sigma_1\#\Sigma_2}$, as in Fig. 7.16.

Theorem 7.6 ([43]). *The bound given in Eq. (7.1) is sharp.*

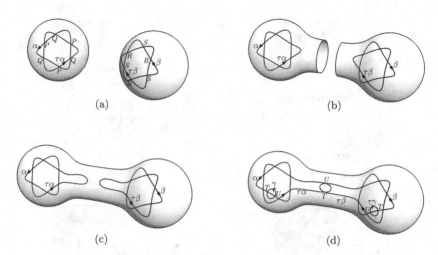

(a)   (b)

(c)   (d)

Fig. 7.16: Surgery between Johansson diagrams.

Example 7.5 is the key in the proof of this theorem. To prove Theorem 7.6, it is enough to show that $t_0\big(L(3,1)\#L(3,1)\big) = 6$. To do so, the first homology and fundamental groups of low Montesinos complexity manifolds must be studied. This study has been done in detail in [43], where the following result is proved:

Theorem 7.7. *If* Σ *is a filling Dehn sphere of a 3-manifold* M *with at most 4 triple points, then* $H_1(M, \mathbb{Z}) \not\cong \mathbb{Z}_3 \oplus \mathbb{Z}_3$.   □

Proof of Theorem 7.6. Let M_1 and M_2 be 3-manifolds. Seifert-van Kampen theorem implies that $\pi_1(M_1\#M_2) = \pi_1(M_1) * \pi_1(M_2)$. Hence

$$\pi_1\big(L(3,1)\#L(3,1)\big) = \mathbb{Z}_3 * \mathbb{Z}_3,$$

and therefore

$$H_1\big(L(3,1)\#L(3,1); \mathbb{Z}\big) = \mathbb{Z}_3 \oplus \mathbb{Z}_3.$$

Theorem 7.7 shows that $t_0\big(L(3,1)\#L(3,1)\big) > 4$. By Lemma 7.1, there is no Dehn sphere with odd number of triple points. Therefore the Dehn sphere in Example 7.5 is minimal, and hence $t_0\big(L(3,1)\#L(3,1)\big) = 6$.   □

The combinatorial properties of the filling Johansson diagrams: (i) connectedness; (ii) even intersection property; and (iii) the *symmetry* between sister curves (when performing a complete travel along sister curves we must pass through the same number of crossings); impose strong restrictions on the diagram groups arising from Johansson diagrams in $\mathbb{S}^2$. These restrictions allow to prove Theorem 7.7 and other results as, for example (cf. Theorem 4.7):

Theorem 7.8 ([43]). *Let Σ be a Dehn sphere with at most two double curves and whose Johansson diagram is connected. The fundamental group of Σ is isomorphic to 1, $\mathbb{Z}$ or $\mathbb{Z}_p$ with $p \le 6$.* $\square$

The following theorem, that is also stated in [43], relies also in purely combinatorial techniques.

Theorem 7.9. *If $t_0(M) \le 4$, the fundamental group of M, if it is not trivial, is isomorphic to either $\mathbb{Z}$, $\mathbb{Z}_q$ with $q \le 6$, $\mathbb{Z} \oplus \mathbb{Z}$ or*
$$K = \langle\, a, b \mid a^2 = b^2 \,\rangle.$$
$\square$

The group K is isomorphic to the fundamental group of the Klein bottle. It is known that no infinite 3-manifold group can be isomorphic to the fundamental group of a closed surface, therefore Theorem 7.9 implies that the fundamental group of a non-simply connected 3-manifold M with $t_0(M) \le 4$ is isomorphic to $\mathbb{Z}$ or $\mathbb{Z}_q$ with $q \le 6$.

7.4 The triple point spectrum

The *genus g triple point number* $t_g(M)$ of the 3-manifold M is defined as the number of triple points of a minimal filling Dehn g-torus of M (see [69]). The ordered sequence of the triple points numbers of M is the *triple point spectrum of M*
$$\mathcal{T}(M) = \big(t_0(M), t_1(M), t_2(M), \ldots\big).$$
If Σ is a filling Dehn g-torus, by Theorem 3.25 it has at least one region. So
$$t_g(M) \ge 2g - 1. \tag{7.2}$$
When $g = 0$, this equation gives a trivial bound. Since $t_0(M) \ge 2$, there is a 'minimal spectrum' for 3-manifolds, which is in fact the spectrum of a well-known manifold:

Theorem 7.10 ([42]). $\mathcal{T}(\mathbb{S}^2 \times \mathbb{S}^1) = (2, 1, 3, 5, 7, 9, \ldots).$ $\square$

Examples in Secs. 7.2.1 and 7.2.2 give the first two elements of the sequence $\mathcal{T}(\mathbb{S}^2 \times \mathbb{S}^1)$. To conclude the proof of Theorem 7.10 it is enough to apply the following lemma.

Lemma 7.11. *For any 3-manifold M and any $g \ge 0$,*
$$t_{g+1}(M) \le t_g(M) + 2.$$

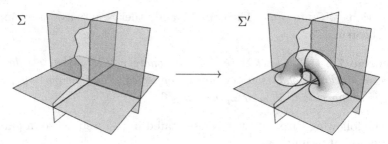

Fig. 7.17: Handle piping.

Proof. Let M be a 3-manifold and let Σ be a minimal Dehn g-torus in M. After modifying a neighbourhood of a triple point of Σ with a *handle piping* as in Fig. 7.17, the new Dehn surface Σ' so obtained is a Dehn $(g+1)$-torus with 2 more triple points than Σ, and by Proposition 2.15 it is still filling. $\qquad\square$

Lemma 7.11 gives an upper bound for the genus g triple point number in terms of the Montesinos complexity of the manifold. The Shima's spheres show that $t_0(\mathbb{S}^3) = 2$. Therefore, Lemma 7.11 and Eq. (7.2) imply that $2g - 1 \le t_g(\mathbb{S}^3) \le 2g + 2$ for any $g \ge 0$.

Theorem 7.12 ([42]). $\mathcal{T}(\mathbb{S}^2) = (2, 4, 6, 8, \ldots)$. $\qquad\square$

This result is a corollary of the following theorem:

Theorem 7.13 ([42]). *If M is a $\mathbb{Z}/2$-homology sphere,*

$$t_g(M) \ge 2g + 2.$$ $\qquad\square$

Theorem 7.4 can be extended to the case of general genus g triple point numbers as follows.

Theorem 7.14 (cf. [43]). *Let M_1 and M_2 be two 3-manifolds, and let $M_1 \# M_2$ be their connected sum. Then*

$$t_{g_1+g_2}(M_1 \# M_2) \le t_{g_1}(M_1) + t_{g_2}(M_2) + 4 \qquad (7.3)$$

for any $g_1, g_2 \ge 0$. If $g_1 = 0$ then

$$t_{g_2}(M_1 \# M_2) \le t_0(M_1) + t_{g_2}(M_2) + 2. \qquad (7.4)$$

Proof. We proceed as in the proof of Theorem 7.4. In fact, it is easy to adapt that proof to the case of Eq. (7.4), since only one the separating face (in one of the surfaces) is needed.

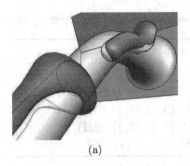
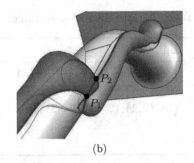

(a)   (b)

Fig. 7.18: Modification of the fingers of Fig. 7.15. (a) The situation corresponding to Fig. 7.15(c), where one of the fingers has been modified to cut the connection along the white tube. (b) The finished modification. Two of the four new triple points appear marked as P_1 and P_2. The other two are in the opposite side of the up-to-down finger of the figure.

For $i = 1, 2$, let Σ_i be a minimal filling Dehn g_i-torus in M_i with p_i triple points and let B_i be an embedded 3-disk in M_i disjoint from Σ_i. The connected sum $M_1 \# M_2$ is obtained from the disjoint union of $M_1 - \text{int}(B_1)$ and $M_2 - \text{int}(B_2)$ when the boundaries of B_1 and B_2 become identified by a homeomorphism. To prove Eq. (7.3), the case when neither Σ_1 nor Σ_2 have separating faces, as the Dehn torus of Sec. 7.2.2, must be also considered. In these cases, there is only one region of $\Sigma_1 \cup \Sigma_2$, and it becomes self-connected by the surgery. To avoid this, the region inside the piping should be cut adding a transverse disk. For example, by replacing the fingers of Fig. 7.15(d) with the ones of Fig. 7.18 it is obtained a filling Dehn $(g_1 + g_2)$-torus in $M_1 \# M_2$ with $p_1 + p_2 + 4$ triple points. $\qquad \square$

7.5 Surface-complexity

In the following paragraphs generic Dehn surfaces are considered with arbitrary domain, perhaps non-connected or non-orientable. A filling (quasi-filling) Dehn surface of the 3-manifold M is *absolutely minimal* if there is no filling (quasi-filling) Dehn surface in M with less triple points than Σ. The triple point numbers are closely related to the *surface-complexity $sc(M)$* of M introduced in [4], defined as the number of triple points of an absolutely minimal quasi-filling Dehn surface of M. Quasi-filling surfaces are more flexible than filling ones, therefore, if the *strong surface-complexity $sc_0(M)$* of M is defined as the number of triple points of an absolutely minimal filling Dehn surface of M, it is clear that $sc(M) \leq sc_0(M)$. Theorem 3 in [4]

Table 7.1: Irreducible orientable 3-manifolds with surface-complexity one.

Elliptic	Euclidean
$L(6,1)$, $L(8,3)$, $L(12,5)$, $L(14,3)$,	$\mathbb{T}^3$,
$\left(\mathbb{S}^2, (2,1), (2,1), (2,-1)\right)$,	$\left(\mathbb{S}^2, (2,1), (4,1), (4,-3)\right)$,
$\left(\mathbb{S}^2, (2,1), (2,1), (3,-2)\right)$,	$\left(\mathbb{S}^2, (3,1), (3,1), (3,-2)\right)$,
	$\left(\mathbb{S}^2, (2,1), (2,1), (2,1), (2,-3)\right)$,
	$\left(\mathbb{RP}^2, (2,1), (2,-1)\right)$.

implies the following surprising result:

Theorem 7.15. *If M is $\mathbb{P}^2$-irreducible and $sc(M) > 0$, then*

$$sc(M) = sc_0(M).$$ $\square$

The $\mathbb{P}^2$-irreducible 3-manifolds with surface-complexity up to 1 have been classified in [4] and [5].

Theorem 7.16 ([4, 5]). *Let M be a $\mathbb{P}^2$-irreducible. Then:*

(i) if $sc(M) = 0$, the manifold M is $\mathbb{S}^3$, $\mathbb{RP}^3$ or $L(4,1)$; and

(ii) if $sc(M) = 1$, then M is one of the manifolds listed in Table 7.1. $\square$

The list up to complexity two has been obtained independently in [38].

Chapter 8

Knots, knots and some open questions

8.1   2-Knots: lifting filling Dehn surfaces

The following question was posed to the first author by R. Fenn:

Question 8.1. *Do filling Dehn spheres in M lift to embeddings in $M \times I$?*

A Dehn surface $\Sigma \subset M$ is *liftable* if there exist a parametrization $f : S \to M$ of Σ and an embedding $\hat{f} : S \to M \times I$ such that $f = p \circ \hat{f}$, where $p : M \times I \to M$ denotes the projection onto the first factor. If there is no such embedding, Σ is *non-liftable*.

Theorem 8.2 ([68]). *Every 3-manifold M has a liftable filling Dehn sphere.*

Proof. Let M be a 3-manifold, let T be a triangulation of M, and take the filling collection spheres Σ_T in M as constructed in Sec. 6.4. Let Σ be a filling Dehn sphere of M obtained by performing spiral pipings around triple points of Σ_T in such a way that there is no spiral piping connecting a vertex sphere with a triangle sphere of Σ_T. Take a copy in $M \times \{0\}$ of all the edge spheres of Σ_T, a copy in $M \times \{1\}$ of all the vertex spheres of Σ_T and a copy in $M \times \{-1\}$ of all the triangle spheres of Σ_T. Let $\xi : \mathbb{S}^1 \times [0, 1] \to M$ be a parametrization of a spiral piping of Σ such that $\xi(\mathbb{S}^1 \times \{0\})$ is contained in an edge sphere of Σ_T. Define $\hat{\xi} : \mathbb{S}^1 \times [0, 1] \to M \times [-1, 1]$ as $\hat{\xi}(z, t) = \big(\xi(z, t), t\big)$ if $\xi(\mathbb{S}^1 \times \{1\})$ is contained in a vertex sphere of Σ_T, or as $\hat{\xi}(z, t) = \big(\xi(z, t), -t\big)$ if $\xi(\mathbb{S}^1 \times \{1\})$ is contained in a triangle sphere of Σ_T. This construction shows that Σ is liftable. $\square$

There are algorithms to determine if a given Dehn surface is liftable or not in terms of its Johansson diagram [12,18]. Using those algorithms, it can be checked that the Shima's spheres, the examples of Secs. 7.2.1 and 7.2.2 and the filling Dehn spheres constructed in [49] and [67] are liftable, while the filling Dehn sphere of Sec. 7.2.3 is non-liftable. In [18], Giller shows that the 2-sphere that runs parallel at both sides of Boy's surface, the *Giller's sphere*, is non-liftable (see also [12]). Giller's sphere is a filling Dehn sphere of $\mathbb{S}^3$, and it is used in [70] to construct a non-liftable filling Dehn sphere in any 3-manifold M. Therefore,

Theorem 8.3 ([70]). *Every 3-manifold M has a non-liftable filling Dehn sphere.*   □

Fenn's question gives a connection between filling Dehn surfaces and 2-knot theory. A liftable Dehn sphere Σ in M is the projection into M of a 2-knot $\hat{\Sigma}$ in $M \times I$. To recover $\hat{\Sigma}$ from Σ some *crossing information* must added to Σ along the double curves of Σ (see [10,12]). This crossing information can be codified as a *colouring* [10] of the Johansson diagram $\mathcal{D}$ of Σ, that is, a map $\sigma : \mathcal{D} \to \{-1, +1\}$ such that $\sigma(\tau\alpha) = -\sigma(\alpha)$ for each $\alpha \in \mathcal{D}$ that encodes which curve of each pair sister curves of $\mathcal{D}$ goes "up" and "down" in $\hat{\Sigma}$. If Σ is filling, it has the advantage that it can be built up in a unique way from its Johansson diagram, and thus $\hat{\Sigma}$ can be built in a unique way from the pair $(\mathcal{D}, \sigma)$, which is a *Johansson representation* of $\hat{\Sigma}$. About that, we believe that it is not difficult to prove the following

Conjecture 8.4. *Every surface-knot $\hat{\Sigma}$ in $M \times I$ can be deformed by isotopy in such a way that $p(\hat{\Sigma})$ is a filling Dehn surface of M.*

This conjecture would imply that every surface-knot has a Johansson representation.

If Σ is non-liftable, it can be pushed into $M \times I$, trying to separate the double curves of Σ using a colouring $\sigma : \mathcal{D} \to \{-1, +1\}$, but the result will be a transversely immersed surface $\hat{\Sigma}$ in $M \times I$ that is embedded except at a finite number of double points, where two sheets of $\hat{\Sigma}$ intersect transversely. The double points of $\hat{\Sigma}$ are lifts of triple points of Σ and, using orientations in the domain surface of Σ and in $M \times I$, each of them can be labelled with a "sign" $+1$ or -1 in a standard way.

We wonder if any of these constructions can be of interest for understanding properties of the 3-manifold M or the 2-knots in $M \times I$. In fact, most of the literature where Dehn surfaces come out appears in the context of 2-knot theory (see for instance [11,12,18,57,62,75]). Perhaps part of the work already done for 2-knots can be adapted to this context to extract

Fig. 8.1: A Dehn sphere containing the trefoil knot.

information about 3-manifolds.

8.2 1-Knots

Filling Dehn surfaces can also be used to study 1-knots, using a construction introduced in [44]. This construction is outlined in the following paragraphs.

Let M be a 3-manifold, and let K be a tame knot or link in M.

Definition 8.5. The filling Dehn surface Σ of M *splits* K if:

(1) K intersects Σ transversely in a finite set of non-singular points of Σ;
(2) $K - \Sigma$ is a disjoint union of open arcs;
(3) for each region R of Σ, if the intersection $R \cap K$ is non-empty it is exactly one unknotted arc in R;
(4) for each face Δ of Σ, the intersection $F \cap K$ contains at most one point.

Theorem 8.6 ([44]). *Every link K in a 3-manifold M can be split by a filling Dehn sphere of M. Moreover, this filling Dehn sphere can be chosen such that it intersects exactly twice each connected component of K.*

Proof. Take a triangulation T of M such that K is simplicial with respect to T. This can be done because K is tame. The filling Dehn surface Σ_T splits K, and any filling Dehn sphere Σ obtained from Σ_T by spiral pipings also splits K.

To prove the second statement, Σ_T is modified slightly around each connected component of K. If K_i is one of such components, instead of introducing one sphere for each vertex and each edge of the triangulation T belonging to K_i, a unique Dehn sphere is taken around K_i as in Figure 8.1, while the rest of Σ_T remains the same. With this modification, Σ_T intersects K_i twice. The desired result is obtained by repeating the same operation around each connected component of K.   □

Assume that the filling Dehn sphere Σ of M splits K. Let $f : \mathbb{S}^2 \to M$ be a parametrization of Σ and let $\mathcal{D}$ be the Johansson diagram of Σ. The pair $\left(\mathcal{D}, f^{-1}(K)\right)$ is a *Johansson diagram* of K. Since M can be built from $\mathcal{D}$ and the arcs of $\Sigma - K$ are unknotted in the regions of Σ:

Proposition 8.7. *It is possible to recover the pair* (M, K) *from a Johansson diagram of* K.   □

Another interesting property is

Proposition 8.8 ([44]). *The groups* $\pi_1(M - K)$ *and* $\pi_1(\Sigma - K)$ *are isomorphic.*   □

Using the presentation of the fundamental group of Dehn surfaces introduced in [42] (Sec. 4.5), this proposition allows to compute the knot group of K from a Johansson diagram of K.

If K is a knot, Σ *diametrically splits* K if it intersects K exactly twice. In this case, Σ provides a *diametral Johansson diagram* of K that has the form $(\mathcal{D}, \{P_1, P_2\})$ for some $P_1, P_2 \in \mathbb{S}^2 - |\mathcal{D}|$.

Splitting filling Dehn surfaces for knots, specially the diametrically ones, are quite useful for studying the branched covers of M. Assume that K is a link in M, and let $r : M^* \to M$ be a branched covering with downstairs branching set K. Let Σ be a filling Dehn surface of M that splits K and take $\Sigma^* = r^{-1}(\Sigma)$.

Theorem 8.9 ([44]). $\Sigma^* = r^{-1}(\Sigma)$ *is a filling Dehn surface of* M^*.   □

Theorem 8.10 ([44]). *If* K *is a knot and* Σ *is a filling Dehn sphere that diametrically splits* K, *then* Σ^* *is a filling collection of spheres in* M, *and* Σ^* *is a filling Dehn sphere of* M *if and only if* r *is locally cyclic.*   □

If the branching set K is a knot, the covering r is *locally cyclic* if the monodromy map sends knot meridians onto n-cycles, where n is the number of sheets of the covering. This is equivalent to say that $r : r^{-1}(K) \to K$ is a homeomorphism. When r is locally cyclic Theorem 8.10 can be used to obtain an algorithm that provides the Johansson diagram of Σ^* from the diametral Johansson diagram of K provided by Σ [44].

8.3 Open problems

Since filling Dehn spheres is a barely explored research topic, there are many natural open questions related to them. Here some of these questions are proposed. The reader may find that some of them are rather ambitious, while others might look "almost trivial" since a good example would solve them.

Let M be a 3-manifold.

8.3.1 *Filling Dehn surfaces and filling Dehn spheres*

It should be interesting to investigate which kind of information about M can be easily obtained from its Johansson representations.

Problem 8.11. *Find 3-manifold invariants that can be computed in an easier way using a Johansson representation of M.*

Any "important subsurface" of M, like separating surfaces, Heegaard surfaces or incompressible surfaces, can be isotoped to be transverse to a given filling Dehn surface, leaving some "footprints" in the Johansson diagram.

Problem 8.12. *Find algorithms to detect "important subsurfaces" of M from a Johansson representation of M.*

A natural problem is also the following:

Problem 8.13. *Write a computer program that obtains the list of all the filling Johansson diagrams in $\mathbb{S}^2$ or in a g-torus with a given number of triple points.*

8.3.2 *Filling homotopies. Moves*

Regarding our main result, we wonder if the proof of Theorem 5.8 can be adapted to a more general context than nulhomotopic filling Dehn spheres.

Conjecture 8.14. *Regularly homotopic filling immersions are f-homotopic.*

It is clear that Theorem 5.8 opens a path for finding new 3-manifold invariants, as it has been done in [3].

Problem 8.15. *Make effective computations of the invariant* inv_m *defined in [3].*

Problem 8.16. *Construct other invariants like* inv_m *following the indications given in [3, Sec.3].*

As it is pointed out in [3], the framework used to define inv_m is analogous to that used to define the Turaev-Viro invariant [65].

Question 8.17 ([3]). *Are the invariant* inv_m *and the Turaev-Viro invariant related (in some sense) to each other?*

Amendola's invariant gives lower bounds for the Matveev complexity of $\mathbb{P}^2$-irreducible 3-manifolds [3].

Question 8.18 ([3]). *Are these lower bounds obtained via* inv_m *sharp, at least for some closed 3-manifolds?*

It would be desirable for the invariants of 3-manifolds constructed from null-homotopic filling Dehn spheres, such as inv_m, to be computable from their Johansson diagrams.

Problem 8.19. *Find a characterization of null-homotopic filling diagrams.*

8.3.3 *Montesinos complexity. Triple point spectrum*

It is not easy to determine if a given filling Dehn surface is minimal or not. Some filling Dehn surfaces among our examples are trivially minimal because they have the minimal possible number of triple points given by the inequalities of Eq. (7.2)

$$t_0(M) \geq 2 \qquad \text{and} \qquad t_g(M) \geq 2g - 1.$$

Until now, the only non-trivial characterization of minimal Dehn surfaces are those given in [43] for a minimal filling Dehn sphere of $L(3,1)\#L(3,1)$ and in [42] for the minimal Dehn g-tori of $\mathbb{S}^3$ with $g > 1$. The argument in both cases is the same:

> *There is a concrete filling Dehn g-torus Σ of M with p triple points, and it is shown that any other Dehn g-torus Σ' with less than p triple points must verify that $H_1(M;\mathbb{Z}) \neq H_1(\Sigma';\mathbb{Z})$.*

This argument is essentially based on the presentation of the fundamental group of a Dehn surface in terms of its Johansson diagram (Chap. 4), and we think that it can only be applied for some concrete examples.

Problem 8.20. *Find, if there is one, a general method to determine if a filling Dehn surface is minimal or not in terms of its Johansson diagram.*

Problem 8.21. *Let M be a 3-manifold. Classify the minimal filling Dehn surfaces of M.*

The Montesinos' and Johansson's spheres are related by a single saddle move (Sec. 7.1). On the other hand, to transform the Banchoff's sphere into other Shima's sphere finger moves +1 are needed. It is possible to check that a saddle move applied to the Banchoff's sphere produces a non-filling result. It is also clear from its Johansson diagram that the Banchoff's sphere does not admit finges moves −1 or −2. Then, to modify the Banchoff's sphere using f-moves, it is always necessary to increase the number of triple points. This suggests the following definition.

Definition 8.22. A filling Dehn surface Σ of a 3-manifold M is *irreducible* if the only f-moves that it admits are finger moves +1 or +2.

Question 8.23. *If a filling Dehn surface is irreducible, is it minimal?*

An affirmative answer to this questions would help to solve Problem 8.20 in some cases, because irreducibility can be checked on the Johansson diagram.

Lemma 7.11 suggests the following definitions.

Definition 8.24. A minimal Dehn g-torus of M is *exceptional* if it has less than $t_{g-1}(M) + 2$ triple points.

Definition 8.25. The *height* $\mathcal{H}(M)$ of M is the highest genus among all exceptional Dehn g-tori of M. If M has no exceptional Dehn g-torus, then $\mathcal{H}(M) = 0$.

Let Σ_g and Σ_{g+1} be minimal filling Dehn g- and $(g + 1)$-tori of M, respectively. By Theorem 3.25, if Σ_{g+1} is exceptional, Σ_{g+1} has less regions than Σ_g. Therefore

Theorem 8.26. *The height of M is finite. In other words, there exists $h \geq 0$ such that $t_{g+1}(M) = t_g(M) + 2$ for all $g \geq h$.* $\qquad\square$

By Sec. 7.4, $\mathcal{H}(\mathbb{S}^3) = 0$ and $\mathcal{H}(\mathbb{S}^2 \times \mathbb{S}^1) = 1$.

Problem 8.27. *Classify the 3-manifolds with small height.*

Question 8.28. *For a given $h \in \{0, 1, 2, \dots\}$, are there a finite number of 3-manifolds with $\mathcal{H}(M) = h$?*

The dual cell decomposition of the one defined by a filling Dehn surface of M is a *cubulation* of M, i.e., a cell decomposition whose building blocks are cubes (see for instance [2], [6] or [17]). In this way, the strong surface-complexity of a 3-manifold M defined in Sec. 7.5 coincides with the minimal number of cubes in a cubulation of M.

Special types of cubulations, and their dual (filling Dehn) surfaces are key tools in some outstanding problems in 3-manifold topology, as in Agol's proof of the virtual Haken conjecture [1], for example. A Dehn g-torus Σ in M with $g > 0$ is *incompressible* if for any parametrization $f : S \to M$ of Σ the induced map $f_* : \pi_1(S) \to \pi_1(M)$ is injective.

Question 8.29. *Do incompressible minimal filling Dehn g-tori exist?*

Or, on the contrary

Question 8.30. *Is an incompressible filling Dehn g-torus in M minimal?*

In Problem 5.13 of Kirby's problem list [39], Habegger proposed a set of moves for cubulations that consist in replacing B with B', where B and B' are complementary balls (union of n-cubes) in the boundary of the standard $(n+1)$-cube. Those moves and a special set of them are called *bubble moves* and *np-bubble moves*, respectively, in [17]. The np-bubble moves are those bubble moves for which B or B' does not contain parallel faces (when viewed in the boundary of the $(n+1)$-cube). Following [17], let denote by $CBB(M)$ the set of equivalence classes of cubulations of M mod np-bubble moves and by $CB(M)$ the set of equivalence classes of cubulations mod bubble moves. In [17] it is proved that $CB(\mathbb{S}^2)$ contains only two equivalence classes.

Problem 8.31. *Compute $CB(\mathbb{S}^3)$ and $CBB(\mathbb{S}^3)$.*

8.3.4　*Knots*

Let K be a knot in M.

Problem 8.32. *Find an algorithm that provides the simplest filling Dehn sphere of M that diametrically splits K.*

Of course, before solving this problem it is necessary to define "simplest", perhaps in terms of the number of triple points and/or double curves of the filling Dehn sphere. Filling Dehn spheres that diametrically split K are useful for studying the branched coverings of M with branching set K. Therefore, if K is a universal knot in $\mathbb{S}^3$ [28] it can be expected to find some kind of "universal family of Johansson diagrams" representing all closed 3-manifolds. Since the simplest universal knot is the figure-eight knot [29]:

Problem 8.33. *Find a filling Dehn sphere in $\mathbb{S}^3$ that diametrically splits the figure-eight knot.*

Appendix A

Proof of Key Lemma 2

The collapsing

Define $D_0 = f(D)$ and $D_1 = g(D)$. The transformation of $f(S)$ into $g(S)$ will be done by a step-by-step deformation of D_0 into D_1. This sequence of deformations will be guided by a collapsing of T_B over D_1. At each intermediate step of the construction, Σ' (resp. f', resp. D') will denote the Dehn surface (resp. immersion, resp. embedded disk) obtained from $f(S)$ (resp. f_T, resp. D_0). In each step of the construction, there will be a closed 3-disk $B' \subset B$ such that Σ' and $g(S)$ are related by the pushing disk (D, B'). The approach will be by repeated application of Propositions 6.37 and 6.38. At each intermediate step, there will be a superregular region R' of Σ' contained in B' with one face Δ' in D', and Σ' is modified by a pushing disk along R' from Δ'. By Proposition 6.38, the resulting Dehn surface is filling and by Proposition 6.37 it can be obtained from Σ' by f-moves supported by D. For simplicity, when this operation is applied, it is said that D' *passes over R'* or that D' *is pushed along R'*, without further mention to superregularity or Propositions 6.37 and 6.38.

Let

$$B \searrow \overset{s_3}{\cdots} \searrow B^2 \searrow \overset{s_2}{\cdots} \searrow B^1 \searrow \overset{s_1}{\cdots} \searrow D_1 \qquad (1)$$

be a sequence of collapsings in decreasing order of dimension (see the proof of Proposition 6.57). That is, the first s_3 collapsings of (1) are 3-collapsings of B and the subcomplex B^2 of B resulting from them is two-dimensional. The next s_2 collapsings of (1) are 2-collapsings of B, and the resulting simplicial complex B^1 is the union of D_1 and a one-dimensional complex. Finally, the last s_1 collapsings of (1) are 1-collapsings and the resulting complex is D_1.

183

Step 1: the 3-collapsings

Step 1.1: the first 3-collapsing

Assume that the first 3-collapsing of (1) is (σ_1, t_1), where σ_1 is a tetrahedron of B and t_1 is a triangle of σ_1 (Fig. A.1). Since this is the first collapsing of the sequence (1), $t_1 \subset D_0$. The triangle t_1 divides the region $R(t_1)$ of Σ_T into two 6-gonal prisms of $f(S)$ one of which, say R_1, is contained in B. Let Δ_1 be the face of R_1 in t_1. Push D along R_1 from Δ_1 (see Figs. A.3, A.4 and A.5).

After this operation it is said that D *has passed over* $R(t_1)$. Note that the new pushing ball B' lying between Σ' and $g(S)$ does not meet $\mathrm{cl}(R(t_1))$.

The new filling Dehn surface Σ' intersects the truncated tetrahedron $R(\sigma_1)$ in an open 6-gon Δ' "parallel" to the top face of R_1. This face Δ' divides $R(\sigma_1)$ into two connected components: one of them is also a truncated tetrahedron $R'(\sigma_1)$, and the other is a 6-gonal prism lying outside B' between $R'(\sigma_1)$ and R_1 (Figs. A.4 and A.6(a)). Push D' along $R'(\sigma_1)$ (Fig. A.6).

At this stage, it is said that D' *has passed over* $R(\sigma_1)$ and that D' *has passed over the* 3-*collapsing* (σ_1, t_1). A schematic view of Step 1.1 is depicted in Fig. A.7.

Step 1.2: the second 3-collapsing

Let Σ' be the Dehn surface obtained at the end of the previous step.

If B is a single tetrahedron of T Step 1 is finished and continue to Step 2. Assume that there is more than one tetrahedron of T in B. Let (σ_2, t_2) the next 3-collapsing in (1).

If t_2 is a triangle of D_0, proceed with (σ_2, t_2) exactly as with (σ_1, t_1).

If t_2 is not a triangle of D_0, it must be one of the triangles of the first tetrahedron σ_1. In this case, there are two possibilities, either $t_2 \subset f(S)$ or $t_2 \not\subset f(S)$. Let Δ' be the intersection of D' with $R(t_2)$, and consider also the open 6-gon $\Delta_2 = t_2 \cap R(t_2)$. These two open disks Δ' and Δ_2 are the bottom and top faces of a 6-gonal prism E_2, respectively.

- If $t_2 \not\subset f(S)$, the surface Σ' can be deformed with an ambient isotopy until D' agrees with the triangle t_2 in a neighbourhood of $\mathrm{cl}(\Delta_2)$.
- If $t_2 \subset f(S)$, push D' along E_2.

In both cases, $t_2 \subset f(S)$ or $t_2 \not\subset f(S)$, after the deformations, D' and $R(\sigma_2)$

are separated by another 6-gonal prism R_2. Finally, proceed exactly as with the first collapsing. After this process, it is said that D' has passed over the 3-collapsing (σ_2, t_2) (see Fig. A.8).

Step 1.3: the remaining 3-collapsings

For the rest of the 3-collapsings of (1), proceed as in Step 1.2 until D' passes over all the 3-collapsings of (1) using f-moves supported by D.

Step 2: 2-collapsings

If e is an edge of T and v_1 and v_2 are the vertices of e, denote by $C(e)$ the closed cylinder which is the closure of $Be - (Bv_1 \cup Bv_2)$. This cylinder $C(e)$ contains $R(e)$ and $R(e, t)$ for each triangle t of T incident with e (Fig. 6.47).

Let $(\tau_1, e_1), \ldots, (\tau_{s_2}, e_{s_2})$ be the 2-collapsings of the sequence (1).

Consider the first 2-collapsing (τ_1, e_1) of the sequence (1) and assume, for simplicity, that τ_1 and e_1 are in the interior of B. In this case, since τ_1 is the unique triangle of B incident with e_1 remaining in the 2-complex B^2, the configuration of Σ' inside the cylinder $C(e_1)$ is the cartesian product of a diagram like the one in Fig. A.9 with a closed interval. Therefore, the configuration of D' around e_1 is similar to the one in Fig. A.10. Moreover, the configuration of D' in a neighbourhood of τ_1 is similar to the one depicted in Fig. A.11(a). Figure A.11(b) is a side view of the scene of Fig. A.11(a), and Figs. A.11(c) to A.11(f) show other point of views of the same configuration, including the 2-spheres of Σ_T meeting D' near τ_1. In a first sub-step (Step 2.1) D' is taken out of the interior of the cylinder $C(e_1)$ (Fig. A.12(b)), and in a subsequent sub-step (Step 2.2) D' is taken out of $R(\tau_1)$ (Figs. A.12(d) and A.12(e)).

Step 2.1: taking D' out of $C(e_1)$

The configuration of Σ' inside the cylinder $C(e_1)$ is like the one in the left-hand side of Fig. A.13, which is just a three-dimensional version of Fig. A.9. If v is one of the two vertices of e_1, an observer located inside Sv would see the picture on the right hand side of Fig. A.13. These figures correspond to the case $e_1 \not\subset f(S)$. Figures A.14 and A.15 show the four possible configurations when $e_1 \subset f(S)$:

(i) $e_1 \not\subset \mathrm{Sing}(f)$ and $\tau_1 \subset f(S)$ (Fig. A.14(a));
(ii) $e_1 \not\subset \mathrm{Sing}(f)$ and $\tau_1 \not\subset f(S)$ (Fig. A.14(b));

(iii) $e_1 \subset \mathrm{Sing}(f)$ and $\tau_1 \subset f(S)$ (Fig. A.15(a)); and
(iv) $e_1 \subset \mathrm{Sing}(f)$ and $\tau_1 \not\subset f(S)$ (Fig. A.15(b)).

Take a triangle $t \neq \tau_1$ incident with e_1. The disk D' meets the 2-gonal prism $R(e_1, t)$ in a 4-gon Δ_t parallel to Se_1. This 4-gon Δ_t cuts $R(e_1, t)$ into a 4-gonal prism outside B' and a 2-gonal prism R_t inside B' slightly smaller than $R(e_1, t)$. Push D' along R_t (Fig. A.16). If the triangle t is contained in $f(S)$, then we can proceed similarly, but now in two steps: first Σ' is pushed along a 3-gonal prism (Fig. A.17(b)); and after that Σ' is pushed along a 2-gonal prism (Fig. A.17(c)). After these modifications the region $R(e_1, t)$ is left outside B'.

The above operation can be repeated for each triangle of T_B incident with e_1 and distinct from τ_1, until the situation shown in Fig. A.18 is reached. Then, after an ambient isotopy, the situation of Fig. A.19 is obtained. Afterwards, D' pass over e_1 to obtain Fig. A.20. If e_1 is not contained in $f(S)$ or if $e_1 \not\subset f(S)$ and $\tau_1 \subset f(S)$ (Fig. A.14(a)), the deformation from Fig. A.19 into Fig. A.20 can be done by an ambient isotopy. In the three remaining cases (corresponding to Figs. A.14(b), A.15(a) and A.15(b)), proceed as indicated in Figs. A.21 to A.23.

Finally, if the situation is like the one of Fig. A.20, pushing D' along a 2-gonal prism (if $\tau_1 \not\subset f(S)$), or along a 3-gonal prism followed by a 2-gonal prism (if $\tau_1 \subset f(S)$) the situation of Fig. A.24 is attained. Repeating this operation, Fig. A.25 is obtained, where the cylinder $C(e_1)$ is left outside of B'. In this situation C' *has passed over* $C(e_1)$.

Step 2.2: getting D' outside $R(\tau_1)$

Now, the situation is the one presented in Fig. A.12(b).

Let v_1^1 and v_1^2 be the two vertices of e_1, and let v_1^3 be the remaining vertex of τ_1. Let e_1^1 and e_1^2 be two edges of τ_1 different from e_1 such that e_1^i incides in v_1^i for $i = 1, 2$. Let us focus on the part of D' inside $R(\tau_1)$. The intersection $D' \cap R(\tau_1)$ is an open 8-gon $\tilde{\Delta}$ with one edge in each of the spheres Sv_1^1 and Sv_1^2 and two edges in each of the spheres Se_1^1, Se_1^2 and Sv_1^3. "Stretch" D' inside $R(\tau_1)$ to make it look flat, the intersection $B' \cap R(\tau_1)$ has the form depicted in Fig. A.12(c), where the vertical sheet corresponds to τ_1. If $\tau_1 \not\subset f(S)$, push D' along the region of Σ' contained in $B' \cap R(\tau_1)$. If $\tau_1 \subset f(S)$, then τ_1 divides $B' \cap R(\tau_1)$ into two symmetric regions of Σ', and D' is pushed consecutively along them. Figure A.12(d) shows the filling Dehn surface obtained. In this situation, it is said that D' *has passed over* $R(\tau_1)$.

The previous techniques apply, with small modifications, to the cases $e_1 \subset D_0, \tau_1 \not\subset D_0$ (Fig. A.26(a)), and $\tau_1 \subset D_0$ (Fig. A.26(b)).

In general, after these operations, it is said that D' *has passed over the 2-collapsing* (τ_1, e_1)

Step 2.3: the remaining 2-collapsings

Now we focus on the second 2-collapsing (τ_2, e_2) of the sequence (1). If the edge e_2 is not incident with τ_1 the configuration of Σ' near τ_2 is similar to the one studied for (τ_1, e_1). If the edge e_2 is incident with τ_1, it is one of the two edges of τ_1 different from e_1 and τ_1 and τ_2 are the only triangles of B^2 incident with e_2. After D' passes over (τ_1, e_1), the configuration of Σ' near τ_2 become similar to the one considered near τ_1 before. In any case, the methods used to make D' pass over (τ_1, e_1) can be used to make it pass over (τ_2, e_2) as well.

The same argument applies for the rest of the 2-collapsings, that is, D' can pass over all the 2-collapsings of (1) using f-moves supported by D.

Step 3: 1-collapsings

Let $(a_1, v_1), \ldots, (a_{s_1}, v_{s_1})$ be the 1-collapsings of (1), and consider the first 1-collapsing (a_1, v_1) of the sequence.

Assume first that v_1 is an interior vertex of B and that $v_1 \notin f(S)$. The edge a_1 is the only edge of B^1 incident with v_1 and $D' \cap Bv_1$ is a 2-sphere parallel to Sv_1 except at $Bv_1 \cap Ba_1$, where $D' \cap Bv_1$ has a "hole" (Fig. A.27).

Now, $D' \cap Bv_1$ is pushed through the interior of Bv_1 until D' meets no 2-sphere of Σ_T inside Bv_1 and outside $Bv_1 \cap Ba_1$.

Consider the triangulation T_{Sv_1} of Sv_1 induced by T. Let $T_{Sv_1}^1$ be the 1-skeleton of T_{Sv_1} and let Υ be a spanning tree of $T_{Sv_1}^1$ (Fig. A.28(e)). The point $\mu_1 = Sv_1 \cap a_1$ is a vertex of T_{Sv_1}. The complementary set $G = Sv_1 - \Upsilon$ is an open disk and any edge of T_{Sv_1} not contained in Υ separates G into two open disks. For each tetrahedron σ of B incident with v_1, there is a region of Σ' in Bv_1 which is a 6-gonal prism having a 6-gonal face in $R(\sigma)$. The saddle move depicted in Fig. A.28(c) connects two of these 6-gonal prisms and also connects different faces of Σ'. Performing similar saddle moves on each edge of $T_{Sv_1}^1 - \Upsilon$ (Fig. A.28(d)), only different regions and faces of Σ' become connected, therefore all those saddle moves are f-moves.

Sometimes, a planar model of the interior of Bv_1 will be useful to understand what is happening inside Bv_1. Consider the following model,

Bv_1 is the upper half-space of the euclidean space $\mathbb{R}^3$ and Sv_1 is the one point compactification of the horizontal plane (see Figs. A.29 to A.33).

Consider now a *leaf* μ of the tree Υ, that is, μ is a degree 1 vertex of Υ. If $\mu \neq \mu_1$, the filling Dehn surface Σ' in a neighbourhood of μ is similar to the one depicted in Fig. A.29. If μ is incident with m edges of T_{Sv_1} $m-1$ saddle moves have been performed over the triangle spheres around μ. Such configuration is called a *flower* of Σ'. Using $m-1$ finger moves -1 and ambient isotopies as in Fig. A.30, the situation depicted in Fig. A.31(b) is obtained. An ambient isotopy leads to Fig. A.31(c), and finally a finger move -1 transforms Σ' into Fig. A.31(d). By Proposition 5.9, all these finger moves -1 are f-moves and at this stage it is said that D' *has passed over the flower at μ*.

Notice that for each saddle in Fig. A.29, the previous modifications (Fig. A.31) only affect the part of the saddle facing towards μ. Therefore, it is possible to do exactly the same transformations starting with Fig. A.32, where D' has already passed over some flowers next to the one depicted.

Assume that D' has passed over the flower at μ as in Fig. A.31. Let Υ' be the tree obtained from Υ after the 1-collapsing of the leaf μ and the edge of Υ incident with μ. If μ' is a leaf of Υ' different from μ_1, the configuration of Σ' in a neighbourhood of μ' is again a flower (Fig. A.33(b)). Therefore, D' can pass over the flower at μ' using f-moves as it was done at μ. Similarly, any sequence of 1-collapsings of Υ has its associated sequence of passings of D' over flowers (Fig. A.33). Since Υ is a tree, Υ collapses over any of its vertices. Considering a collapsing of Υ over μ_1 and the corresponding sequence of passings of D' over flowers, when Υ collapses over μ_1 the configuration of $D' \cap Bv$, up to ambient isotopy, is like the "lightbulb" of Fig. A.34(d).

If the vertex v_1 of B belongs to $f(S)$, the construction to get the lightbulb configuration of Fig. A.34(d) is a bit longer. Figure A.35 shows the eight possible situations that can occur in this case, depending on the relative positions of v_1 and a_1 with respect to $f(S)$ and $\text{Sing}(f)$: if v_1 is a simple point of f, the edge a_1 can be contained in $f(S)$ (Fig. A.35(d)) or not (Fig. A.35(e)); if v_1 is a double or triple point of f, then $a_1 \not\subset f(S)$ (Figs. A.35(b) and A.35(f)), $a_1 \subset f(S) - \text{Sing}(f)$ (Figs. A.35(c) and A.35(g)), or $a_1 \subset \text{Sing}(f)$ (Figs. A.35(a) and A.35(h)).

Since f is simplicial with respect to T, the intersection $f(S) \cap Sv_1$ is a subcomplex Λ of $T^1_{Sv_1}$. Additionally, $Sv_1 - \Lambda$ is the union of 2, 4 or 8 open disks depending on whether v_1 is a simple, double or a triple point of $f(S)$, respectively.

As in the previous case, consider a spanning tree Υ of $T^1_{Sv_1}$. Again, a saddle move like the one in Fig. A.28(c) on each edge of $T^1_{Sv_1} - \Upsilon$ connects different regions and faces of Σ'. Therefore, the resulting Dehn surface of M is still filling. When these saddle moves are applied over edges of Λ, they are now singular saddle moves $+1$.

Let μ be a leaf of Υ different from μ_1, and let α be the edge of Υ incident with μ. There are two possibilities: $\alpha \subset f(S)$ (Fig. A.36) and $\alpha \not\subset f(S)$ (Fig. A.40).

If $\alpha \subset f(S)$, the situation near μ is like one of those depicted in Fig. A.36. In the situation of Fig. A.36(a), using some finger moves -1 and three finger moves -2 Fig. A.38(a) is obtained. A singular saddle move $+1$ over the edge α produces Fig. A.38(b). Figure A.39(a) is obtained after a finger move -2, and another finger move -2 produces Fig. A.39(b). Replacing some finger moves -2 with f-finger moves -1, this construction applies to the configuration of Fig. A.36(b).

In any case, if $\alpha \subset f(S)$, D' can pass over the flower at μ using f-moves supported by D.

If $\alpha \not\subset f(S)$, the situation is one of the two depicted in Fig. A.40. In Fig. A.40(a) μ is a double point of f. Using finger moves -2 and f-finger moves -1 as in the previous cases, Fig. A.41(a) is obtained. The saddle move of Fig. A.41(b) is an f-move if and only if the faces Δ' and Δ'' of Σ' in $D' \cap Bv_1$ lying at both sides of α (Fig. A.40(a)) agree. In this case, it is said that μ is a *bad leaf* of Υ. Otherwise it is a *good leaf*. If μ is not a bad leaf, the saddle move of Fig. A.41(b) followed by a finger move -1 (Fig. A.41(c)) and a finger move -2 (a finger move -1 if we were in the situation of Fig. A.40(b)) take us to the scene depicted in Fig. A.41(d), when it is said that D' has passed over the flower at μ.

Thus, D' can pass over any flower located at a leaf of Υ different from μ_1 that is not a bad leaf (using f-moves supported by D). This operation is parallel to the collapsing (α, μ) of the tree Υ, after which a subtree Υ' of Υ is obtained. Choosing a good leaf μ' of Υ' different of μ_1 that is not a bad leaf the process can be repeated, making D' pass over the flower at μ' using f-moves supported by D and, in parallel, making Υ' collapse onto a subtree.

Repeating this operation as many times as possible, a situation is reached in which D' has passed over all the flowers at the vertices of T_{Sv_1} except those located in a subtree Υ^* of Υ whose leaves different from μ_1 are all bad leaves. If μ^* is a bad leaf of Υ^*, α^* is the edge of Υ^* incident with μ^*, and L is the connected component of $Sv_1 - \Lambda$ containing α^*, $L - \Upsilon^*$ must be connected because otherwise the saddle move of Fig. A.41(b) would be an f-move at μ^*. This implies that μ^* and μ_1 are the only leaves of Υ^* and that $\mu_1 \notin \Lambda$.

Therefore, if $\mu_1 \in \Lambda$, then $\Upsilon^* = \{\mu_1\}$ and D' does not intersect any 2-sphere of Σ_T, except in a neighbourhood of μ_1. Recall that the initial configuration of D' and $f(S)$ inside Bv_1 was as in Figs. A.35(a), A.35(c), A.35(d), A.35(g) or A.35(h). Then, after ambient isotopy (Figs. A.35(a) and A.35(d)), or after an f-finger move -1 (Fig. A.35(c)), an f-finger move $-3/2$ (Fig. A.35(g)), or a finger move -2 (Fig. A.35(h)), together with ambient isotopies, the lightbulb situation of Fig. A.34(d) is obtained.

If $\mu_1 \notin \Lambda$, at this stage, the situation inside Bv_1 is like the one depicted in Fig. A.43(a). With an ambient isotopy, the part of D' inside Bv_1 not meeting Σ_T is pushed (Fig. A.43) until $D' \cap Bv_1$ looks like Fig. A.44(b), where the part of $D' \cap Bv_1$ that goes parallel to Sv_1 corresponds to the intersection of Σ_T with $D' \cap Bv_1$ over Υ^* and the drop-like part at the centre of the ball goes parallel to $f(S)$. The intersection of D' with $f(S)$ near Bv_1 is like one of those depicted in Fig. A.45 (recall that the initial configuration of D' and $f(S)$ inside Bv_1 was as depicted in Figs. A.35(b), A.35(e), or A.35(f)). Using ambient isotopy (Figs. Fig. A.45(a) and A.45(b)), an f-finger move -1 (Fig. A.45(c)), a finger move -2 (Fig. A.45(d)) or an f-finger move $-3/2$ (Fig. A.45(e)), we make $D' \cap Bv_1$ look like in Fig. A.44(c), which is a *snake* lying over Υ^* with its "head" at μ^* and its "tail" at μ_1. See Figs. A.46 and A.47. Since μ^* can be a simple or double point of f, these two possibilities are depicted in these and subsequent figures.

Figure A.48 shows a closer look at snake's head. In particular, Figs. A.48(e) and A.48(f) shows the snake's head making its top part invisible. An f-finger move -1 transforms Fig. A.49(a) into Fig. A.49(b). In these two figures $f(S)$ has not been included to have a clearer view. If there were a sheet of $f(S)$ transverse to this move, it should be replaced by a finger move -2. A side view of the same operation appears in Figs. A.49(c) and A.49(d). Repeating this operation as many times as needed, Fig. A.49(f) is obtained (Fig. A.49(e) shows a side view). After introducing the sheets of $f(S)$, this picture looks like Figs. A.49(g) or A.49(h). By ambient isotopy (Figs. A.50(a), A.50(b) and A.50(c)) we get Figs. A.50(d) or A.50(e). An f-finger move $-3/2$ (former case) or -1 (latter case), turns both pictures into Fig. A.50(f), and after a last f-finger move -1, Fig. A.50(g) is obtained.

All these operations allow to pass from Figs. A.46(a) and A.46(b) (side views in Figs. A.47(b) and A.47(c), respectively) to Figs. A.51(c) and A.51(d), respectively (side views in Figs. A.51(a) and A.51(b)), only using f-moves supported by D. The operation is repeated, now without sheets of $f(S)$ intersecting the snake, as many times as needed until snake's head is over μ_1. At this moment $D' \cap Bv_1$ is the lightbulb of Fig. A.34(d) (see also Fig. A.51(f)). Figure A.52(a) shows a better view of the interior, by making part of the lightbulb invisible.

If t is a triangle of T incident with a_1, $R(v_1, t) \cap B'$ contains one or two regions of Σ': one trihedron if $t \not\subset f(S)$ (Fig. A.52(b)); or two tetrahedra if $t \subset f(S)$ (Fig. A.53(a)). In the former case, pushing D' along the trihedron Fig. A.52(c) is obtained. In the latter case, a singular saddle move (Fig. A.53(b)) and a finger move -2 lead to Fig. A.53(c). In both cases, it is said that D' has passed over $R(v_1, t)$. If $a_1 \subset f(S)$, this operation can be performed for every triangle of T incident with a_1 without taking care about fillingness since the surface obtained after each of these operations is always filling (Fig. A.54(h)). After that, an $\boldsymbol{f}$-finger move -1 or a finger move -2 leads to Fig. A.54(f), where $D' \cap Bv_1$ is completely contained in Ba_1 and it is said that D' has passed over $R(v_1)$. If $a_1 \not\subset f(S)$ and D' passes over $R(v_1, t)$ for every triangle t of T incident with a_1, the resulting surface is non-filling (Fig. A.54(d)). If there are n_1 such triangles, this operation can be performed at most $n_1 - 1$ times without loosing fillingness (Fig. A.54(b)). After that ambient isotopy and an $\boldsymbol{f}$-finger move -1 take us to the situation of Fig. A.54(f), where D' has passed over $R(v_1)$. In both cases, $a_1 \subset f(S)$ and $a_1 \not\subset f(S)$, D' passes over $R(v_1)$ only using $\boldsymbol{f}$-moves supported by D. After ambient isotopy, $D' \cap Bv_1$ looks like in Fig. A.54(g).

A similar operation as above must be done for making D' pass over $R(v_1, a_1)$ (this case is illustrated in Fig. A.55). After that, by ambient isotopy $D' \cap Ba_1$ is pushed through Bv_1' (Fig. A.56), where v_1' is the opposite vertex of v_1 in a_1, obtaining a configuration for $D' \cap C(a_1)$ as the one depicted in Fig. A.56(f). This configuration is exactly the same as that of Fig. A.55(a), which was also equivalent to that of Fig. A.51(f). Repeating again the operations of the previous paragraph, D' pass over all the regions of Σ' contained in $C(a_1)$ (Fig. A.56(g)). At this stage, it is said that D' *has passed over the 1-collapsing* (v_1, a_1).

In the previous construction it was assumed that v_1 was interior to B. The case $v_1 \in D_0$ must be also considered, that has also two subcases: $a_1 \not\subset D_0$ (Fig. A.57(a)) and $a_1 \subset D_0$ (Fig. A.57(b)). Above techniques, with slight modifications, apply also to this case.

After Step 2, D' looks in a neighbourhood of v_1 like in Fig. A.58 (if $a_1 \not\subset D_0$) or in Fig. A.60 (if $a_1 \subset D_0$). For these two subcases if $v_1 \in f(S-D)$

the possible intersections of $f(S - D)$ with D' inside Bv_1 are depicted in Figs. A.59 and A.61.

We proceed exactly as in the case $v_1 \in \text{int}(B)$ but, instead of the triangulation T_{Sv_1} induced by T in Sv_1, the 2-disk $Dv_1 = Sv_1 \cap B$ and the triangulation T_{Dv_1} that T induces on it must be considered. Take a maximal tree Υ of the 1-skeleton $T^1_{Dv_1}$ and perform a saddle move over each edge of $T^1_{Dv_1} - \Upsilon$. Passing over flowers following a collapsing of Υ into $\mu_1 = Sv_1 \cap a_1$, we make D' pass over all the regions of Σ' contained in regions of the form $R(v_1, \cdot)$ or $R(v_1, \cdot, \cdot)$ except those adjacent to $Sa_1 \cap B'$. The main difference with the case $v_1 \in \text{int}(B')$ is that it will appear *half flowers* as those depicted in Figs. A.62(a) and A.63(b). But it is easy to check that half flowers can be passed over by D' exactly as complete flowers, until a situation as that of Fig. A.62(d) is obtained (see also Fig. A.63(c)). If $f(S-D)$ intersects $D' \cap Bv_1$ as in Figs. A.59(b), A.59(c), A.61(c) or A.61(d), it would be also necessary to apply the "snake trick" of the previous case, perhaps with a "half snake's head", but the previous arguments apply here as well.

Thus, in the case $v_1 \in D_0$ a lightbulb like that of Fig. A.34(d) can be obtained only using f-moves supported by f. If $a_1 \not\subset D_0$, this situation would be exactly the same as in the case $v_1 \in \text{int}(B)$. Therefore, it is possible to proceed exactly as in that case.

If $a_1 \subset D_0$ a *half lightbulb* as the one depicted in Fig. A.64(a) is obtained, intersected by a sheet of $f(S - D)$ if $a_1 \subset f(S - D)$ (Fig. A.64(b)). The methods of the case $v_1 \in \text{int}(B)$ can be applied here to introduce $D' \cap Bv_1$ into Ba_1 (first) and to obtain the situation of Fig. A.64(c), where D' has passed over (the regions of Σ' contained in) $R(v_1)$. After that, by similar modifications as those of Figs. A.55 and A.56, $D' \cap Ba_1$ is completely introduced into Bv'_1, where v'_1 is the opposite vertex of v_1 in a_1. Therefore, D' can pass over the 1-collapsing (a_1, v_1) only using f-moves supported by D as in the case $v_1 \in \text{int}(B)$.

If (a_2, v_2) is the next 1-collapsing in the sequence (1), the situation inside Bv_2 is exactly as one of those studied for Bv_1. Therefore, using the same techniques as above D' passes over the 1-collapsing (a_2, v_2) using f-moves supported by D. Inductively, D' passes over all the 1-collapsing of (1) only using f-moves supported by D (Figs. A.66 and A.67).

Step 4: end of the proof

After Step 3, D' is quite close to D_1. Indeed, it is contained in:

$$\bigcup_{\kappa \in T, \kappa \subset D_1} B\kappa.$$

If τ is a triangle of T contained in D_1, the intersection $D' \cap R(\tau)$ has not changed since Step 1. The set $D' \cap R(\tau)$ is an open 6-gon parallel to $\tau \cap R(\tau)$. If τ is the triangle depicted in Fig. A.68 and it is assumed that B is located under τ in that figure, the 2-disk D' looks near τ like in Fig. A.69(a), or Fig. A.69(b) if the vertex spheres are added to the same picture. A closer view of the same scene, also adding part of $S\tau$ and the edge spheres around τ, appears in Fig. A.70(a). It is clear that Σ' can be modified by ambient isotopy until it coincides with τ inside $R(\tau)$ (Fig. A.70(b)). Repeat this operation for all the triangles of T in D_1.

Let a be an edge of T contained in D_1. The intersection $D' \cap C(a)$ has not changed since Step 2. The 2-disk D' has a configuration inside Ba similar to the one depicted in Fig. A.71(a), if $a \subset \text{int}(D_1)$, or in Fig. A.72(c), if $a \subset \text{eq}(B)$.

Assume first that a is like in Fig. A.71(a), i.e., $a \subset \text{int}(D_1)$. Passing D' along a 2-gonal prism, Fig. A.71(b) is obtained. Repeating the same operation for each triangle sphere intersecting D' inside Ba we arrive to Fig. A.71(c) and, after ambient isotopy, to Fig. A.71(d), where D' coincides with D_1 inside $C(a)$. If $a \subset f(S - D)$, there is a sheet of $f(S - D)$ bisecting one of those 2-gonal prisms into two 3-gonal prism (Fig. A.71(e)). In this case, the passing of D' over these 3-gonal prisms is decomposed into a pushing of D' along one of them followed by a pushing of D' along a 2-gonal prism (see Fig. A.17).

Assume now that $a \subset \text{eq}(B)$. Since (D, B) is strongly transverse, it is impossible that $a \subset \text{Sing}(f)$. So there is no other sheet of $f(S)$ intersecting a except the one containing D_0. We are in the situation of Fig. A.72(c). After pushing D' along a 2-gonal prism for each triangle sphere intersecting $C(a) \cap B$, Fig. A.72(e) is obtained. After ambient isotopy Fig. A.72(f) is got, where D' coincides with D_1 inside $C(a)$.

Repeat above operations for each edge of T contained in D_1.

Finally, if v is a vertex of D_1, $D' \cap Bv$ looks like in Fig. A.74, if $v \in \text{int}(D_1)$, or Figs. A.79(b) and A.79(c), if $v \in \text{eq}(B)$. In the former case, there might be one or two sheets of $f(S - D)$ intersecting Bv (Fig. A.78), while in the latter case there might be one sheet of $f(S - D)$ intersecting Bv (Fig. A.79(d)). The "passing-over-flowers" operations introduced in Step 3, with slight modifications in the case $v \in \mathrm{eq}(B)$, can be applied to these cases in order to make D' pass over all the regions of Σ' contained in the 3-disks $B\epsilon$ with $\epsilon \in T_B$, $\epsilon \not\subset D_1$ that are still contained in B'. After that, by ambient isotopy we make D' coincide with D_1 inside Bv. These constructions are depicted in Figs. A.75, A.76, A.77, and A.79. Repeating this operation for all $v \in D_1$, a filling immersion f' that agrees with g_T except at D is obtained by f-moves and ambient isotopies supported by D. By Remark 6.3, f' and g_T are ambient isotopic in their domain by an ambient isotopy supported by D. $\square$

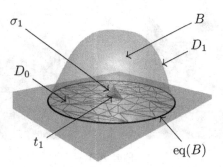

Fig. A.1

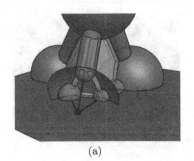

(a)

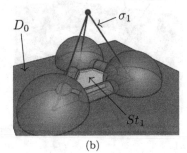

(b)

Fig. A.2

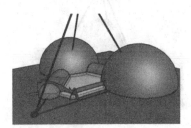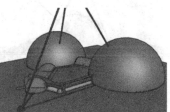

Fig. A.3

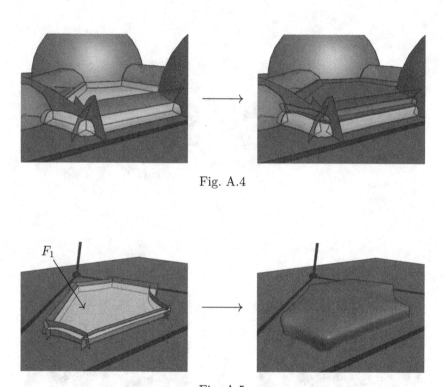

Fig. A.4

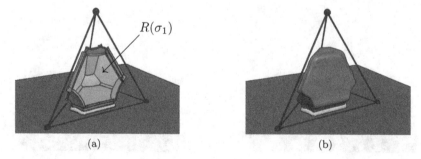

Fig. A.5

Fig. A.6

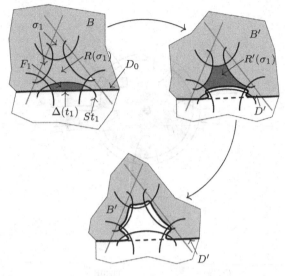

Fig. A.7

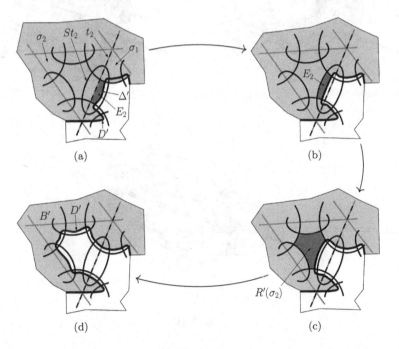

Fig. A.8

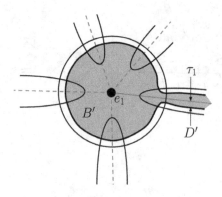

Fig. A.9

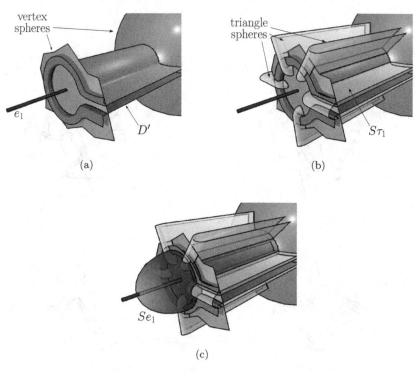

(a)

(b)

(c)

Fig. A.10

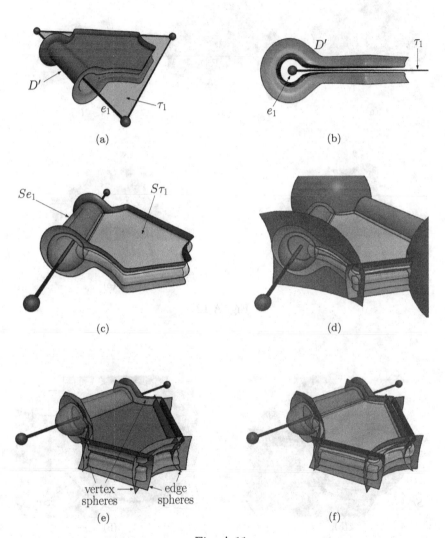

Fig. A.11

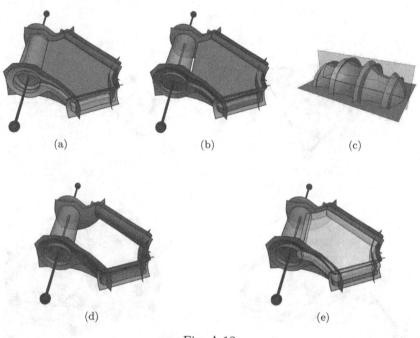

(a)   (b)   (c)

(d)   (e)

Fig. A.12

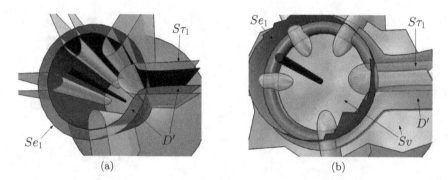

(a)   (b)

Fig. A.13

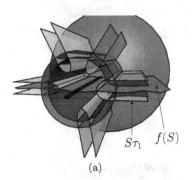 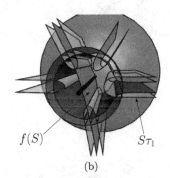

Fig. A.14

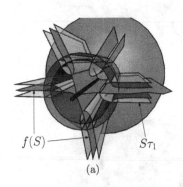 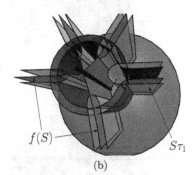

Fig. A.15

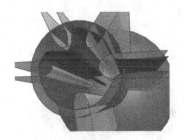 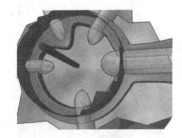

Fig. A.16

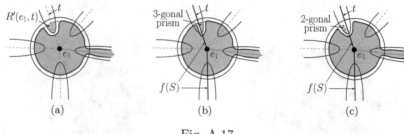

Fig. A.17

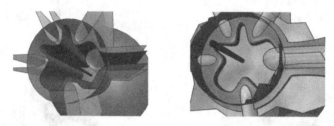

Fig. A.18

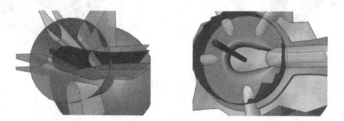

Fig. A.19

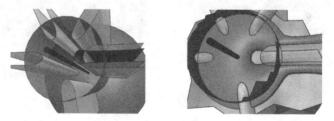

Fig. A.20

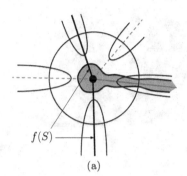 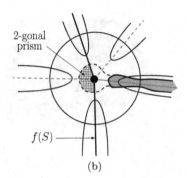

Fig. A.21

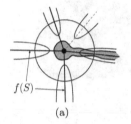 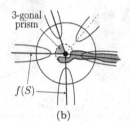 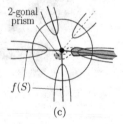

Fig. A.22

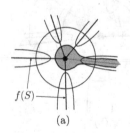 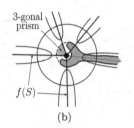 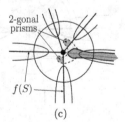

Fig. A.23

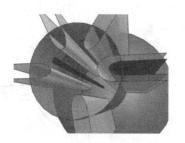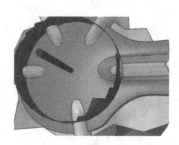

Fig. A.24

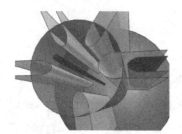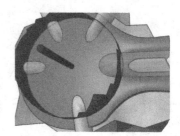

Fig. A.25

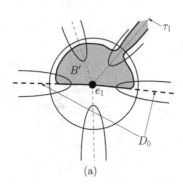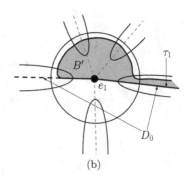

(a)   (b)

Fig. A.26

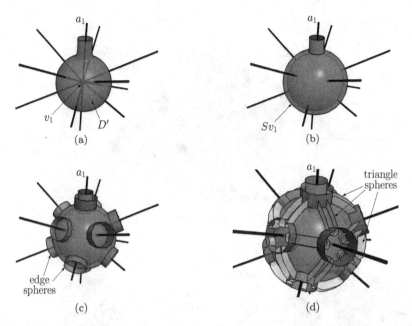

Fig. A.27

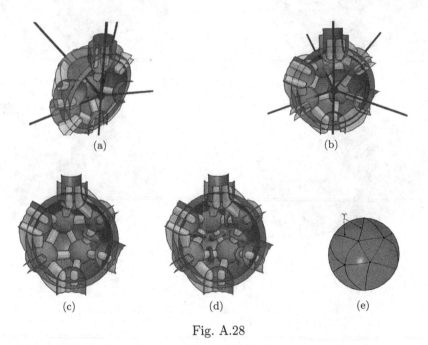

Fig. A.28

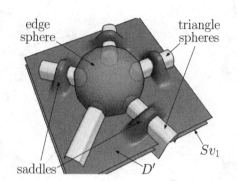

Fig. A.29

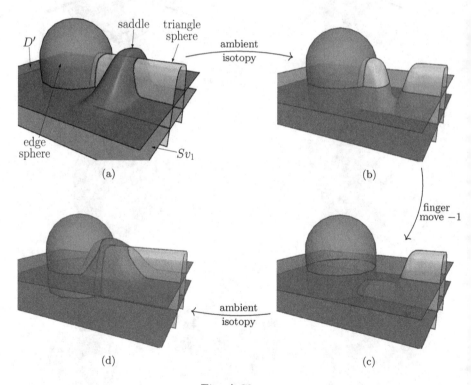

Fig. A.30

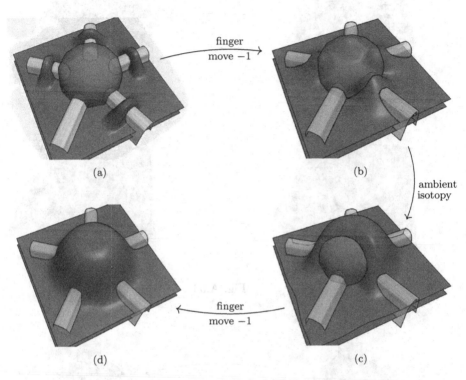

Fig. A.31

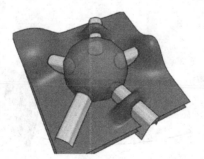

Fig. A.32

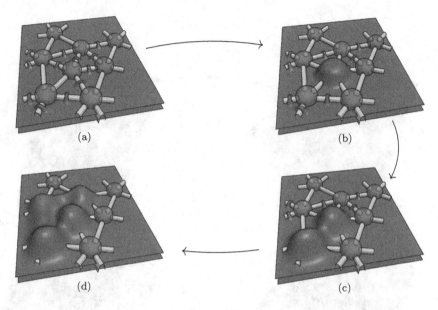

(a)

(b)

(d)

(c)

Fig. A.33

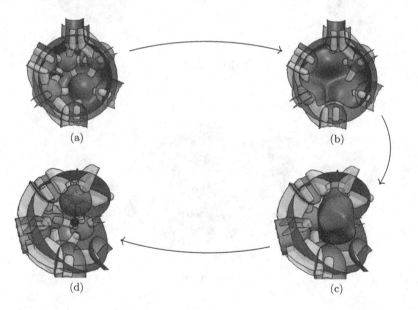

(a)

(b)

(d)

(c)

Fig. A.34

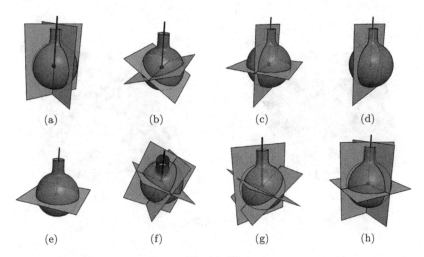

Fig. A.35

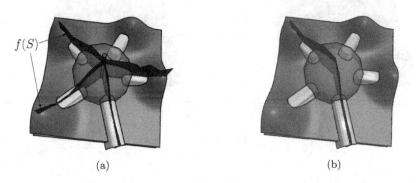

Fig. A.36

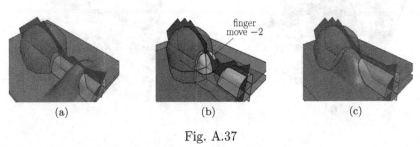

Fig. A.37

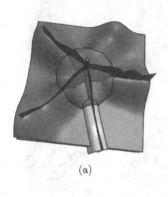

(a)

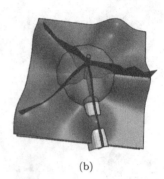

(b)

Fig. A.38

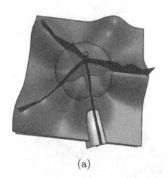

(a)

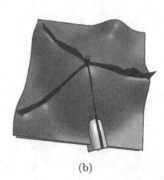

(b)

Fig. A.39

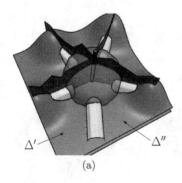

(a)

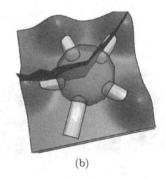

(b)

Fig. A.40

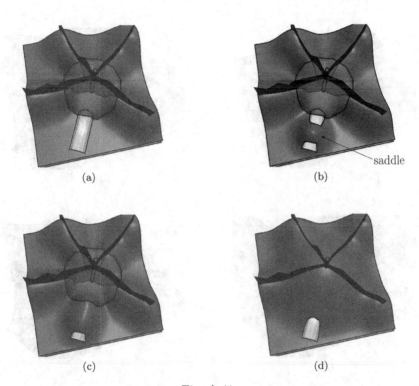

(a)

(b)

saddle

(c)

(d)

Fig. A.41

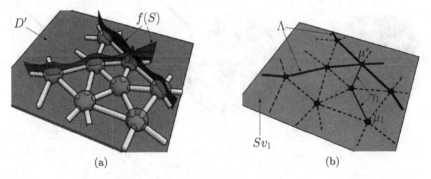

D'

$f(S)$

Λ

μ_2

η

μ_1

Sv_1

(a)

(b)

Fig. A.42

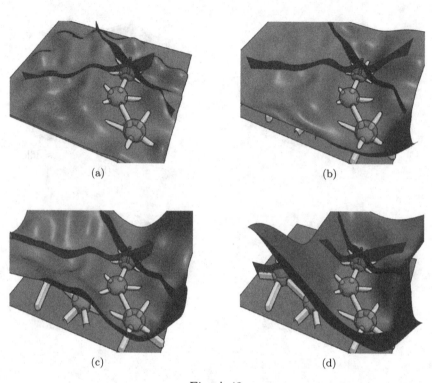

(a)

(b)

(c)

(d)

Fig. A.43

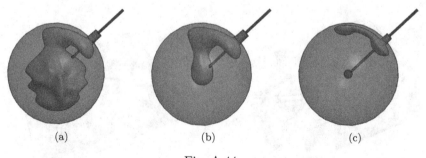

(a)

(b)

(c)

Fig. A.44

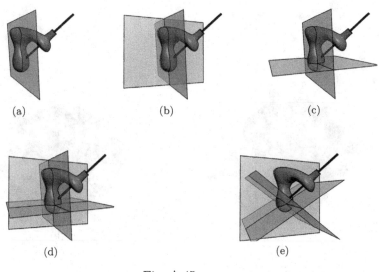

(a)   (b)   (c)

(d)   (e)

Fig. A.45

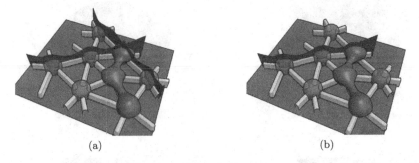

(a)   (b)

Fig. A.46

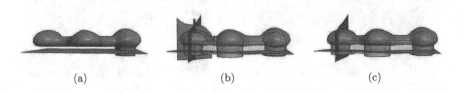

(a)   (b)   (c)

Fig. A.47

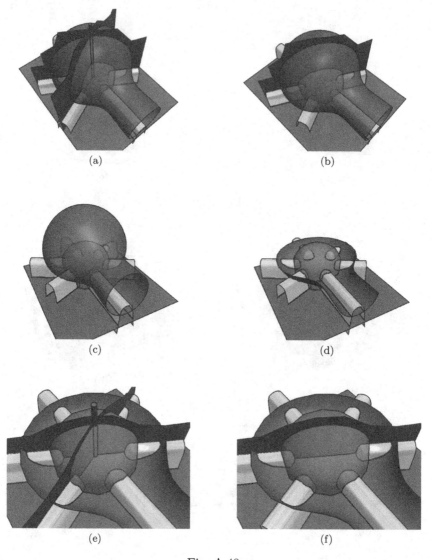

(a)   (b)

(c)   (d)

(e)   (f)

Fig. A.48

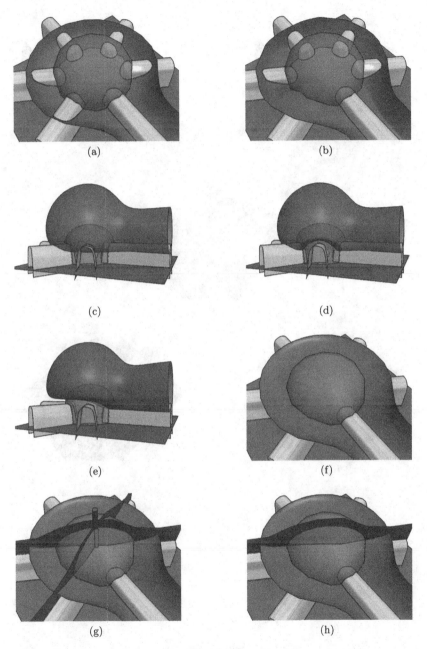

(a)　　　　　(b)

(c)　　　　　(d)

(e)　　　　　(f)

(g)　　　　　(h)

Fig. A.49

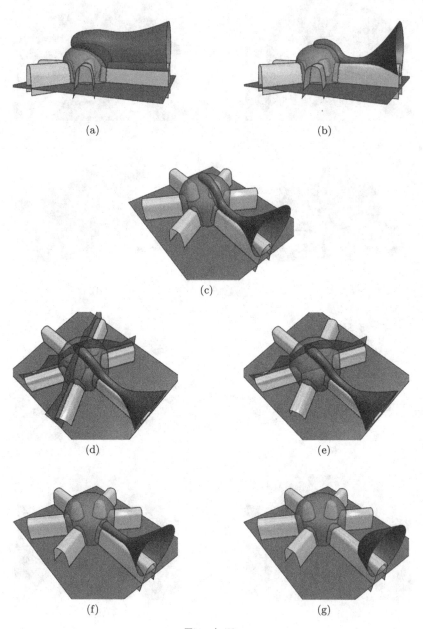

(a)

(b)

(c)

(d)

(e)

(f)

(g)

Fig. A.50

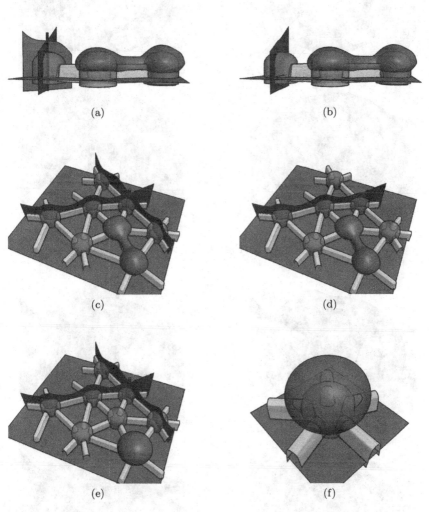

(a)

(b)

(c)

(d)

(e)

(f)

Fig. A.51

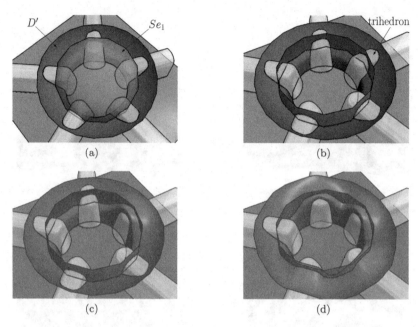

Fig. A.52

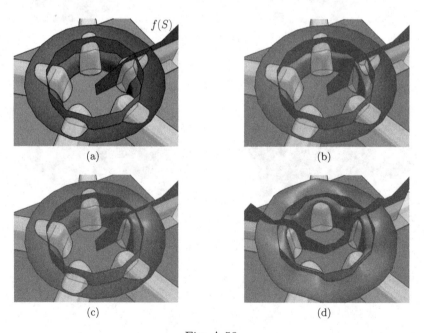

Fig. A.53

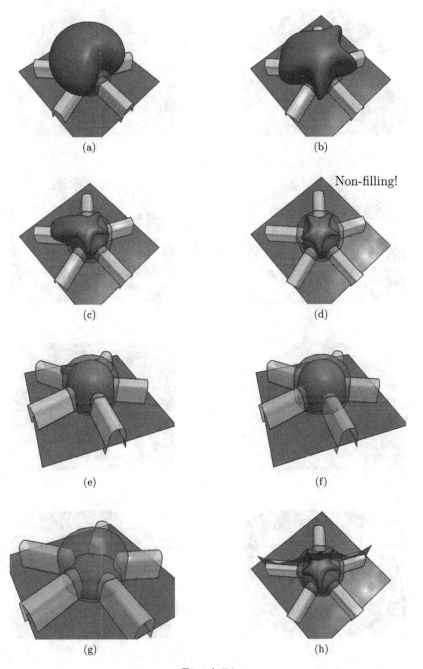

Non-filling!

(a)   (b)

(c)   (d)

(e)   (f)

(g)   (h)

Fig. A.54

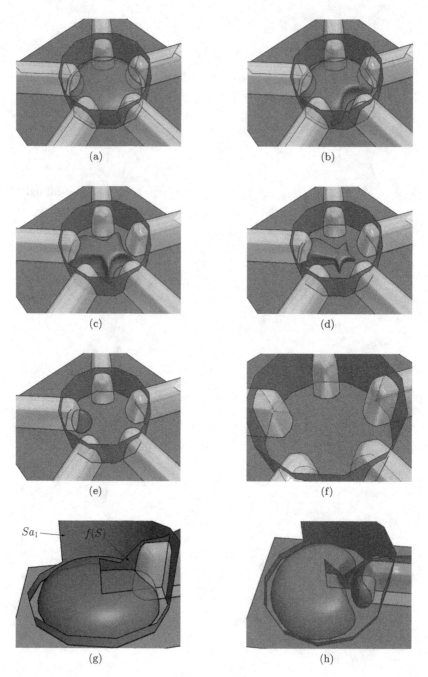

(a)   (b)

(c)   (d)

(e)   (f)

(g)   (h)

Fig. A.55

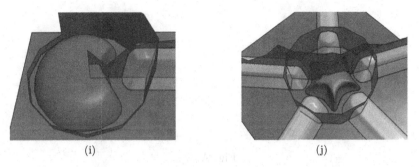

(i) (j)

Fig. A.55: (cont.)

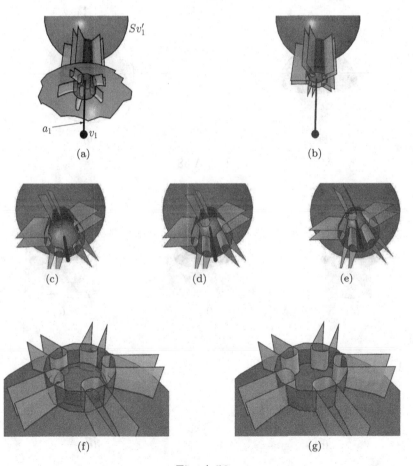

(a) (b)

(c) (d) (e)

(f) (g)

Fig. A.56

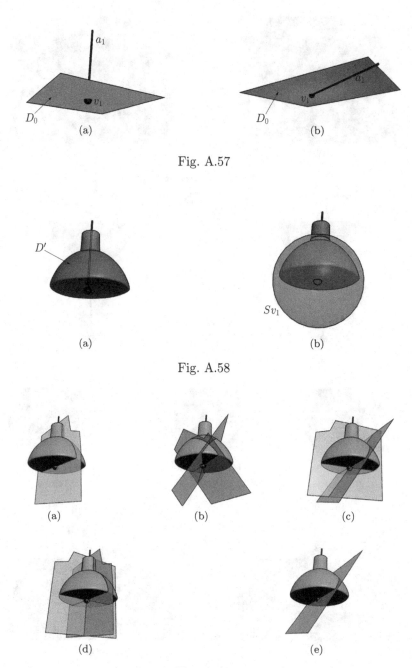

Fig. A.57

Fig. A.58

Fig. A.59

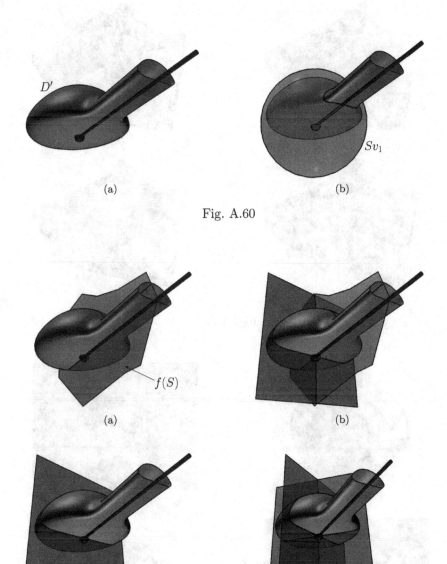

Fig. A.60

Fig. A.61

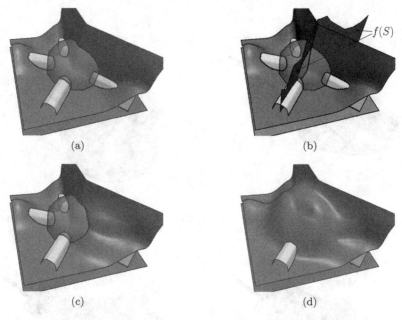

Fig. A.62

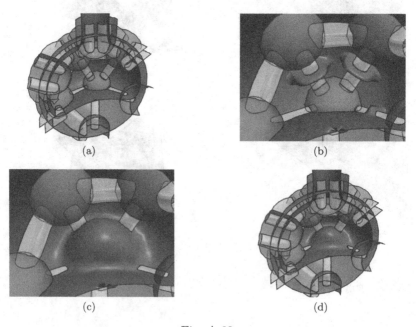

Fig. A.63

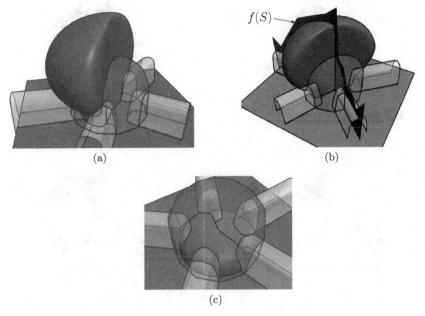

Fig. A.64

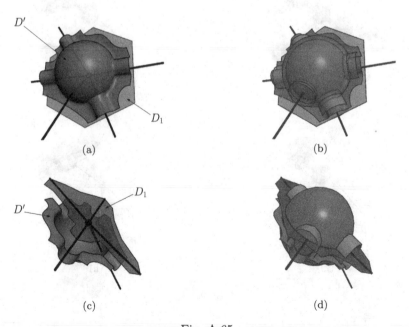

Fig. A.65

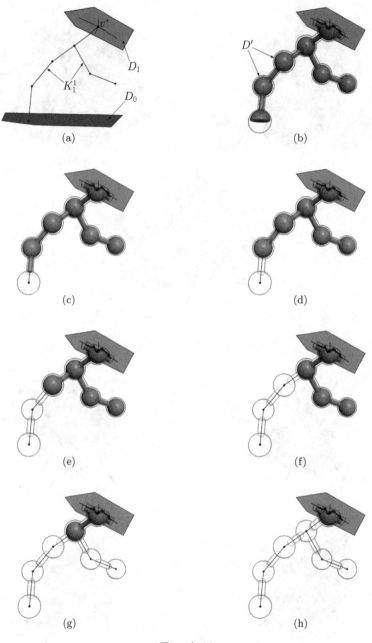

Fig. A.66

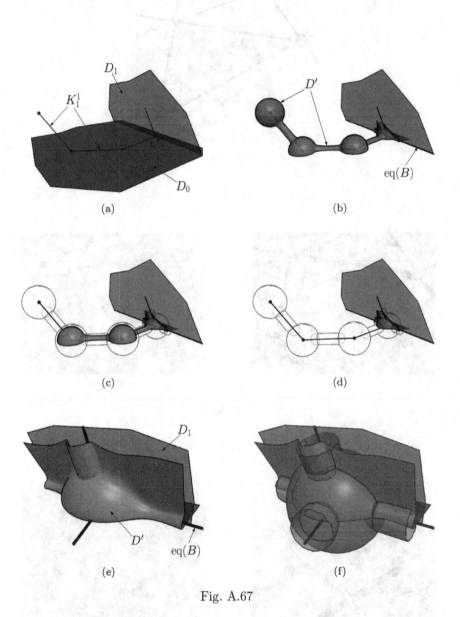

Fig. A.67

Fig. A.68

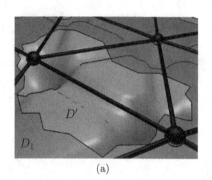

(a)

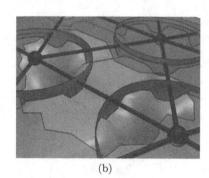

(b)

Fig. A.69

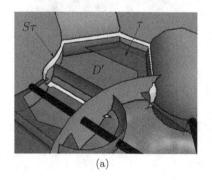

(a)

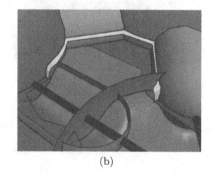

(b)

Fig. A.70

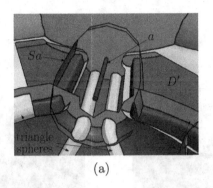

(a)

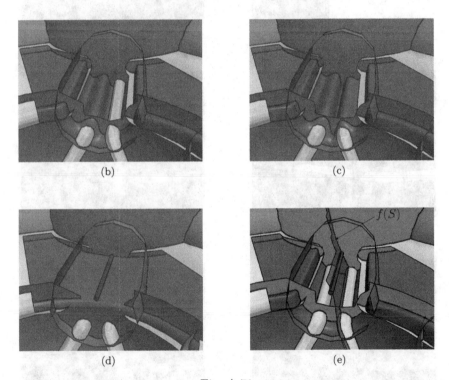

(b)   (c)

(d)   (e)

Fig. A.71

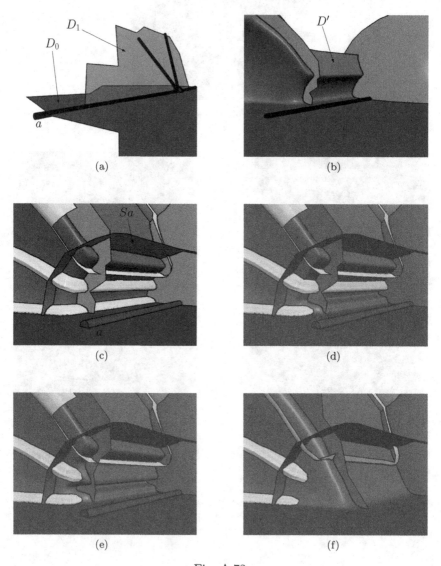

Fig. A.72

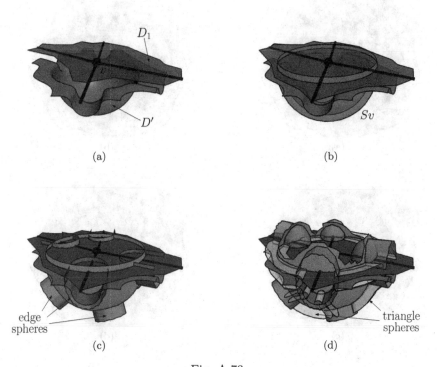

Fig. A.73

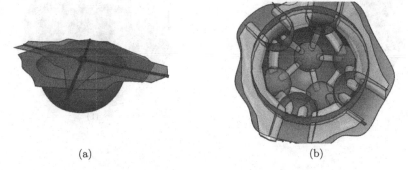

Fig. A.74

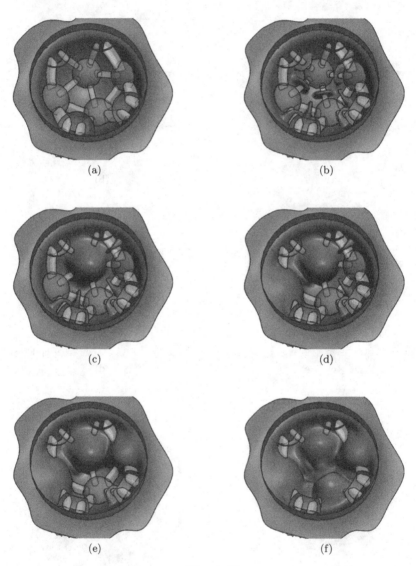

(a)   (b)

(c)   (d)

(e)   (f)

Fig. A.75

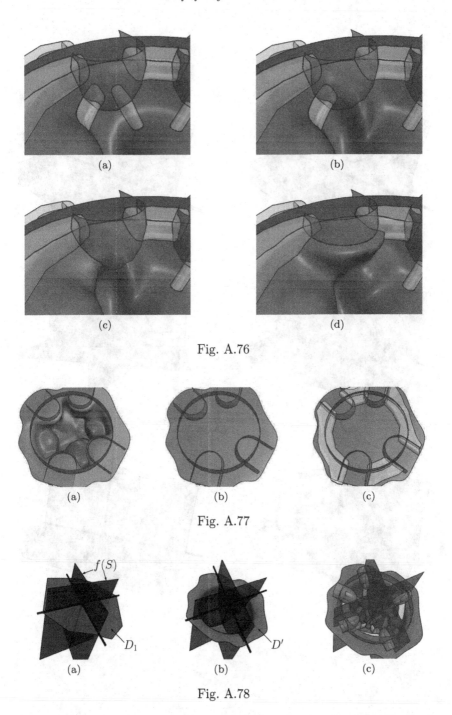

Fig. A.76

Fig. A.77

Fig. A.78

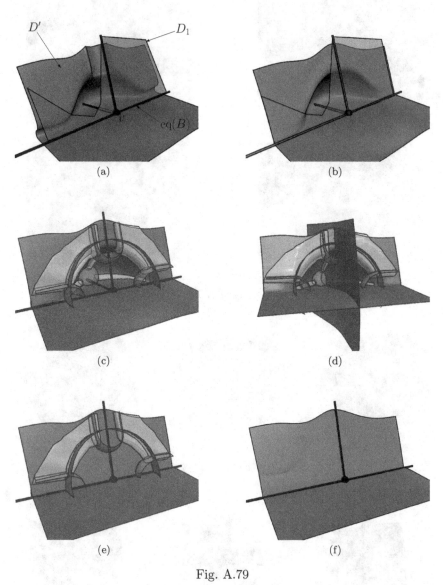

Fig. A.79

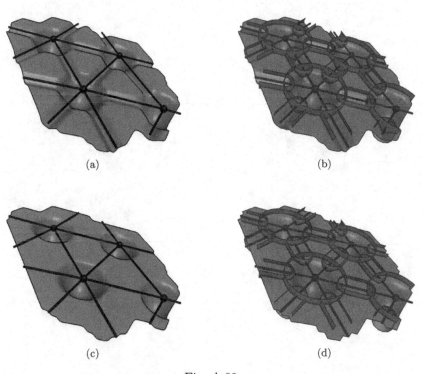

(a)

(b)

(c)

(d)

Fig. A.80

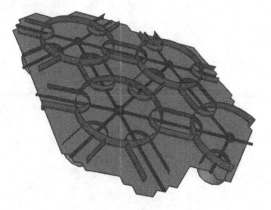

Fig. A.81

Appendix B

Proof of Lemma 6.46

In both cases (a) and (b), it is straightforward to see that $\Sigma \cup \Sigma_{(K \cup \mathrm{cl}(t))}$ fills M.

For a clearer notation, Σ' denotes the Dehn surface recursively constructed at each step of the construction. Then, at a first step $\Sigma' = \Sigma \cup \Sigma_K$.

In the following, if A and B are two Dehn surfaces such that $A \cup B$ is again a Dehn surface, $A \# B$ denotes the Dehn surface obtained from $A \cup B$ after a spiral piping connecting A with B. For the cases (a) and (b) we consider the following subcases:

(a.1) t has no edges in K,

(a.2) t has exactly one edge in K,

(a.3) t has two edges in K,

(a.4) t has its three edges in K,

(b.1) $\mathrm{cl}(t) \cap K$ is the closure of a single edge of t,

(b.2) $\mathrm{cl}(t) \cap K$ is the closure of two edges of t, and

(b.3) $\mathrm{cl}(t) \cap K$ is the full boundary ∂t of t.

Assume that t has vertices v_1, v_2, v_3 and edges e_1, e_2, e_3 with $\partial e_1 = \{v_1, v_2\}$, $\partial e_2 = \{v_2, v_3\}$ and $\partial e_3 = \{v_3, v_1\}$.

Case a.1. The triangle t could have any of its edges in $\mathrm{Sing}(\Sigma)$ or any of its vertices in K (Fig. B.1). If $v_i \in K$ with $i = 1, 2, 3$, the vertex sphere Sv_i is contained in $\Sigma \cup \Sigma_K$ (Fig. B.1(b)). If t has vertices in K we assume that $v_1 \in K$. If t has no vertices in K, assume that $v_1 \in \mathrm{Sing}(\Sigma)$.

Step 1: inflating Sv_1 from Σ. If $v_1 \in K$, then $Sv_1 \subset \Sigma'$: bypass this step and continue to Step 2.

If $v_1 \notin K$, there is one edge e of T incident with v_1 and contained in

237

Sing(Σ). If any of the edges of t incident with v_1 is contained in Sing(Σ) it is possible to assume that $e = e_1$ (this is the case depicted in Fig. B.2). Consider the point Q_0 of intersection of Sv_1 with e (Fig. B.2). The point Q_0 is a double point of Σ'. Inflate Q_0 (see Sec. 6.3.5) to obtain a small 2-sphere $\Sigma_{Q_0} \subset Bv_1$ connected with Σ' by a spiral piping as in Fig. B.3(a). Let $\Sigma'\#\Sigma_{Q_0}$ be the resulting surface. If v_1 is not a triple point of Σ, an ambient isotopy transforms $\Sigma'\#\Sigma_{Q_0}$ into $\Sigma'\#Sv_1$. If v_1 is a triple point of Σ, a finger move $+2$ through v_1 (Fig. B.3(b)) and ambient isotopy transform $\Sigma'\#\Sigma_{Q_0}$ into $\Sigma'\#Sv_1$.

Step 2: inflating Se_1 from Σ and Sv_1. Let Q_1 be one of the two points of $\Sigma \cap Sv_1 \cap Se_1$. The point Q_1 is now a double point of Σ'. Inflate Q_1 by modifying Sv_1 in order to obtain a 2-sphere Σ_{Q_1} connected with Sv_1 by a spiral piping.

If $e_1 \subset$ Sing(Σ), then $Q_0 \in e_1$. In this case, after a piping passing move through Q_0 and:

(i) ambient isotopy, if $v_2 \notin K$; or
(ii) a finger move $+2$ and ambient isotopy, if $v_2 \in K$ (Fig. B.5);

the surface $\Sigma'\#Se_1$ is obtained (Fig. B.4).

If $e_1 \not\subset$ Sing(Σ), $\Sigma'\#Se_1$ is obtained after

(i) an ambient isotopy, if $v_2 \notin K$; or
(ii) after a finger move $+1$ and ambient isotopy if $v_2 \in K$.

The surface Σ has been modified only in a 2-disk near Q_0, and the spiral pipings performed around Q_0 and Q_1 are the only ones that will be carried out around points of Σ.

Step 3: inflating St from Sv_1 and Se_1. Let P_1 be one of the two points of $Sv_1 \cap Se_1 \cap St$ (Fig. B.6). The point P_1 is now a double point of Σ'. Inflate it to obtain $\Sigma'\#\Sigma_{P_1}$, with Σ_{P_1} contained in Bt_1 (Fig. B.7(a)). After a finger move $+2$ (Fig. B.7(b)), the 2-sphere Σ_{P_1} crosses t.

• If none of the vertices v_2 and v_3 of t belong to K, $\Sigma'\#\Sigma_{P_1}$ can be transformed into $\Sigma'\#St$ by ambient isotopy (Fig. B.6).
• If $v_2 \in K$, then $Sv_2 \subset \Sigma'$. A finger move $+2$ should be used to make Σ_{P_1} cross Sv_2 (Fig. B.7(c)).
• If $v_3 \in K$, finger move $+1$ is needed to make Σ_{P_1} cross Sv_3 (Fig. B.7(d)).

In each case, after these deformations there exists an ambient isotopy transforming $\Sigma'\#\Sigma_{P_1}$ into $\Sigma'\#St$.

Step 4: inflating Sv_2 from St_1 and Se_1. Consider now the vertex v_2 of e_1 different from v_1. If $v_2 \in K$, then $Sv_2 \subset \Sigma'$ and it is possible to continue to Step 5. Therefore assume that $v_2 \notin K$.

Let P_2 be a point of $Sv_2 \cap Se_1 \cap St_1$. This point is a double point of Σ' (Fig. B.8(a)): inflate it to obtain a small 2-sphere Σ_{P_2} contained in Bv_1 and connected with Σ' by a spiral piping. Figure B.8(b) results from a finger move $+2$.

If $e_1 \subset \mathrm{Sing}(\Sigma)$, another finger move $+2$ produces Fig. B.8(c) and, an ambient isotopy (if $v_2 \notin T(\Sigma)$) or a finger move $+2$ followed by an ambient isotopy (if $v_2 \in T(\Sigma)$ Fig. B.8(d)) gives $\Sigma'\#Sv_2$ (Fig. B.9).

If $e_1 \not\subset \mathrm{Sing}(\Sigma)$, an ambient isotopy (if $v_2 \notin \mathrm{Sing}(\Sigma)$), or a finger move $+1$ and an ambient isotopy (if $v_2 \in \mathrm{Sing}(\Sigma) - T(\Sigma)$), or a finger move $+3/2$ followed by ambient isotopy (if $v_2 \in T(\Sigma)$) gives $\Sigma'\#Sv_2$.

Step 5: inflating Se_2 from St and Sv_2. Take a point P_3 in the set $Sv_2 \cap St \cap Se_2$, and inflate it to obtain $\Sigma'\#\Sigma_{P_3}$. After a finger move $+2$, Σ_{P_3} crosses t (Fig. B.10(a)).

If $e_2 \not\subset \mathrm{Sing}(\Sigma)$ and $v_3 \notin K$, the Dehn surfaces $\Sigma'\#\Sigma_{P_3}$ and $\Sigma'\#Se_2$ are ambient isotopic (Fig. B.11(a)).

If $e_2 \subset \mathrm{Sing}(\Sigma)$ and $v_3 \notin K$, the edge e_2 is contained in a double curve of Σ. Hence, another finger move $+2$ should be applied to make Σ_{P_3} cross the sheet of Σ "orthogonal" to t along e_2 (Fig. B.10(b)). After that, by ambient isotopy $\Sigma'\#\Sigma_{P_3}$ is transformed into $\Sigma'\#Se_2$.

If $v_3 \in K$ and $e_2 \not\subset \mathrm{Sing}(\Sigma)$, a finger move $+2$ makes Σ_{P_3} cross Sv_3 (Fig. B.10(c)). After this move $\Sigma'\#\Sigma_{P_3}$ and $\Sigma'\#Se_2$ are ambient isotopic.

If $e_2 \subset \mathrm{Sing}(\Sigma)$ and $v_3 \in K$, it is possible to pass from Fig. B.10(a) to Fig. B.10(d) by pushing $\Sigma'\#\Sigma_{P_3}$ along a 5-gon inside $\Sigma'\#\Sigma_{P_3}$. By Proposition 6.36, this operation can be decomposed into f-moves. Finally, an ambient isotopy deforms $\Sigma'\#\Sigma_{P_3}$ into $\Sigma'\#Se_2$.

Step 6: inflating Sv_3 from St and Se_2. If v_3 is contained in K, Sv_3 is contained in Σ' and there is nothing to do. In this case, continue to Step 7. Hence, assume that $v_3 \notin K$.

Consider a point P_4 in the intersection $Se_2 \cap St \cap Sv_3$. This is a double point of Σ': inflate it to obtain the filling Dehn surface $\Sigma'\#\Sigma_{P_4}$. As in the previous cases, apply a finger move $+2$ to make Σ_{P_4} cross t (Fig. B.11(b)). After that, several possibilities arise.

If $v_3 \notin \mathrm{Sing}(\Sigma)$, the Dehn surface $\Sigma'\#\Sigma_{P_4}$ is ambient isotopic to $\Sigma'\#Sv_3$ (Fig. B.12(a)).

If $e_2 \subset \mathrm{Sing}(\Sigma)$, apply a finger move $+2$ to make Σ_{P_4} cross the sheet of Σ orthogonal to t along the edge e_2 (Fig. B.11(c)). If v_3 is not a triple point of Σ, the resulting surface is ambient isotopic to $\Sigma'\#Sv_3$. If v_3 is a triple point of Σ, another finger move $+2$ is needed to pass through v_3 (Fig. B.11(d)) and a last ambient isotopy to transform $\Sigma'\#\Sigma_{P_4}$ into $\Sigma'\#Sv_3$.

If $e_2 \notin \mathrm{Sing}(\Sigma)$ but v_3 is a double point of Σ, apply a finger move $+1$ to make Σ_{P_4} cross the sheet of Σ orthogonal to t in v_3 (Fig. B.11(e)). Afterwards, there exists an ambient isotopy transforming $\Sigma'\#\Sigma_{P_4}$ into $\Sigma'\#Sv_3$.

If $e_2 \notin \mathrm{Sing}(\Sigma)$ but v_3 is a triple point of Σ, after a finger move $+3/2$ through v_3, a surface ambient isotopic to $\Sigma'\#Sv_3$ is obtained (Fig. B.11(f)).

Step 7: inflating Se_3 from St and Sv_3. If e_3 is the remaining edge of t, let P_5 be a point of $Sv_3 \cap St \cap Se_3$. This point is a double point of Σ' and inflate it to obtain $\Sigma'\#\Sigma_{P_5}$. After a finger move $+2$, Σ_{P_5} crosses t (Fig. B.12(b)). Now, the same situation is like the one of Step 5 when the vertex v_3 belongs to K.

If $e_3 \not\subset \mathrm{Sing}(\Sigma)$, a finger move $+2$ move parallel to e_3 makes Σ_{P_5} cross Sv_1 (Fig. B.12(c)) and then, by an ambient isotopy, $\Sigma'\#Sv_3$ is obtained.

If $e_3 \subset \mathrm{Sing}(\Sigma)$, pushing $\Sigma'\#\Sigma_{P_5}$ along a 5-gon inside $\Sigma'\#\Sigma_{P_5}$ (Fig. B.12(d)), a Dehn surface ambient isotopic to $\Sigma'\#Sv_3$ is obtained.

After Steps 1 to 7 we have inflated $\mathrm{cl}(t)$ from $\Sigma \cup \Sigma_K$ (Fig. B.13).

Case a.2. Assume that t has exactly one edge in K. Ignoring the sheets of Σ transverse to the one containing t, near t, the surface $\Sigma \cup \Sigma_K$ is like the Dehn surface depicted in Fig. B.14(a) (if $v_3 \notin K$) or Fig. B.14(b) (if $v_3 \in K$). In this case, it is possible to proceed exactly as was done in Case a.1, Step 3, when $v_2 \in K$ (Fig. B.7), to inflate St from Sv_1 and Se_1. Next, proceed as in Case a.1, Steps 4 to 7.

Case a.3. If t has two edges in K, then the situation around t is like that depicted in Fig. B.15(a). As in Step 3, consider a point P_1 in $Sv_1 \cap Se_1 \cap St$. Inflate P_1 and apply to the resulting 2-sphere Σ_{P_1} a sequence of four finger moves $+2$ to get a picture similar to the one in Fig. B.15(c), which is ambient isotopic to $\Sigma' \# St$ (Fig. B.15(d)). After that, apply Case a.1, Step 7, without modification.

Case a.4. If t has its three edges in K, the situation around t is like in Fig. B.16(a). Consider the point P_1 as in Case a.1, Step 3. Inflate P_1 to get the 2-sphere Σ_{P_1} connected with Σ', and apply a finger move $+2$ to make Σ_{P_1} cross t (Fig. B.16(b)). Afterwards, pushing Σ_{P_1} along a 7-gon inside $\Sigma' \# \Sigma_{P_1}$ we get a surface ambient isotopic to $\Sigma' \# St$ (Fig. B.16(c)).

This concludes the proof of Case (a).

In Case (b), the intersection $\mathrm{cl}(t) \cap K$ should agree with the closure of one, two or the three edges of t. For the proof of this case we will also use of Figs. B.1 to B.16(c), but we must remember that, in this case, the sheet of Σ containing t depicted in those figures no longer exists in the surface $\Sigma \cup \Sigma_K$. This implies that some of the moves involved in the proof of Case (a) must be reinterpreted. In particular: "vertical" finger moves $+2$, along which the small spheres Σ_{P_i} crossed t, are now ambient isotopies; "horizontal" finger moves $+2$, parallel to t, are now finger moves $+1$; and pushing disks along n-gons inside Σ are now pushing disks along n-gons outside Σ.

Case b.1. If $\mathrm{cl}(t) \cap K$ is the closure of a single edge of t, we can assume that $\mathrm{cl}(t) \cap K = \mathrm{cl}(e_1)$. Moreover, if only one of the remaining edges e_2 and e_3 of t is contained in Σ, we can assume that this edge is e_2. The situation in a neighbourhood of t is like in Fig. B.14(a). Proceed as in the case $t \subset \Sigma$ with the reinterpretation of moves mentioned above:

(1) Inflating P_1 as in Fig. B.7(a), after a finger move $+1$ (Fig. B.7(c)) and ambient isotopy, St is inflated from Σ'.
(2) Inflating P_3 as in Fig. B.10(a), after: an ambient isotopy (if $e_2 \not\subset \Sigma$); a finger move $+1$ followed by ambient isotopy (if $e_2 \subset \Sigma - \mathrm{Sing}(\Sigma)$, Fig. B.10(b)); or a finger move $+3/2$ followed by ambient isotopy (if $e_2 \subset \mathrm{Sing}(\Sigma)$); Se_2 is inflated from Σ'.
(3) Inflating P_4 as in Fig. B.11(b), after: an ambient isotopy (if $e_2 \not\subset \Sigma$); a finger move $+1$ followed by ambient isotopy (if $e_2 \subset \Sigma - \mathrm{Sing}(\Sigma), v_3 \notin$

Sing(Σ), Fig. B.11(c)); two finger moves $+1$ followed by ambient isotopy (if $e_2 \subset \Sigma - \text{Sing}(\Sigma)$, $v_3 \in \text{Sing}(\Sigma) - \text{T}(\Sigma)$); a finger move $+1$ followed by a finger move $+3/2$ and ambient isotopy (if $e_2 \subset \Sigma - \text{Sing}(\Sigma)$ and $v_3 \in \text{T}(\Sigma)$); a finger move $+3/2$ followed by ambient isotopy (if $e_2 \subset \text{Sing}(\Sigma)$ and $v_3 \in \text{Sing}(\Sigma) - \text{T}(\Sigma)$); or a finger move $+3/2$ followed by a finger move $+2$ and ambient isotopy (if $e_2 \subset \text{Sing}(\Sigma)$ and $v_3 \in \text{T}(\Sigma)$); Sv_3 is inflated from Σ'.

(4) Inflating P_5 as in Fig. B.12(b), after: a finger move $+1$ followed by ambient isotopy (if $e_3 \not\subset \Sigma$, Fig. B.12(c)); or a pushing disk along a 5-gon outside Σ' (if $e_3 \subset \Sigma - \text{Sing}(\Sigma)$, Fig. B.12(d)) followed by ambient isotopy; or a finger move $+3/2$ followed by a finger move $+2$ along e_3 and a pushing disk along a 3-gon outside Σ' (if $e_3 \subset \text{Sing}(\Sigma)$); Se_3 is inflated from Σ'.

After those steps, $\text{cl}(t)$ has been inflated from $\Sigma \cup \Sigma_K$.

Case b.2. If $\text{cl}(t) \cap K$ is the closure of two edges of t, then $\Sigma' = \Sigma \cup \Sigma_K$ meets t as in Fig. B.15(a). Inflate P_1 as in Fig. B.15(b), and Fig. B.15(c) is obtained after three consecutive finger moves $+1$. After an ambient isotopy, we finish to inflate St from Σ'. Next, proceed as in the last step of Case b.2 to inflate Se_3 from Σ'.

Case b.3. If $\text{cl}(t) \cap K$ is the full boundary ∂t of t (Fig. B.16(a)), inflate P_1 (Fig. B.16(b)), and after a pushing disk along a 7-gon outside Σ', $\Sigma' \# St$ is obtained (Fig. B.16(c)).

This finishes the proof.   □

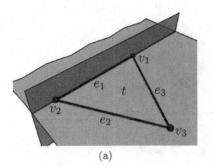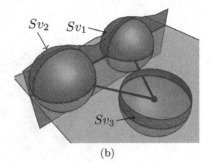

(a)   (b)

Fig. B.1

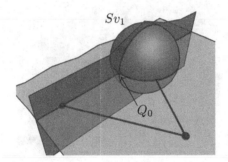

Fig. B.2

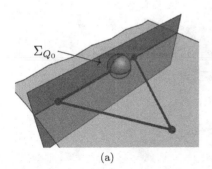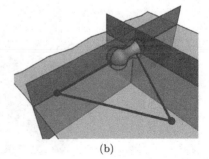

(a)   (b)

Fig. B.3

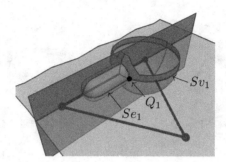

Fig. B.4

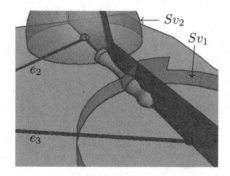

Fig. B.5

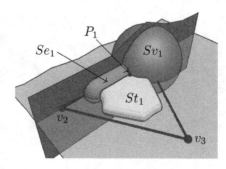

Fig. B.6

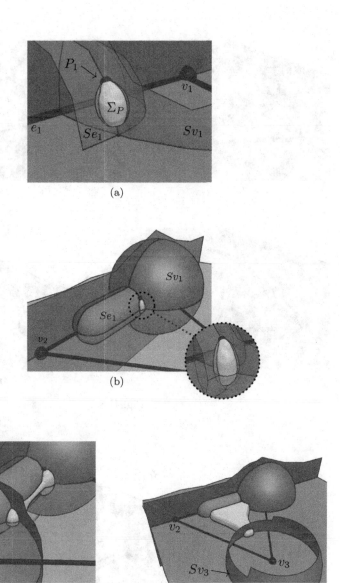

Fig. B.7

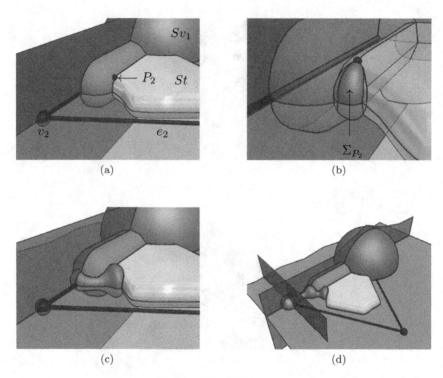

(a)   (b)

(c)   (d)

Fig. B.8

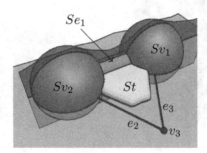

Fig. B.9

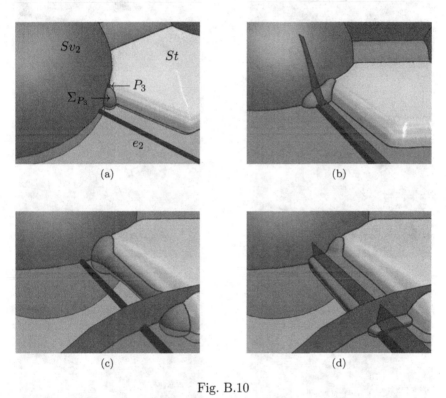

Fig. B.10

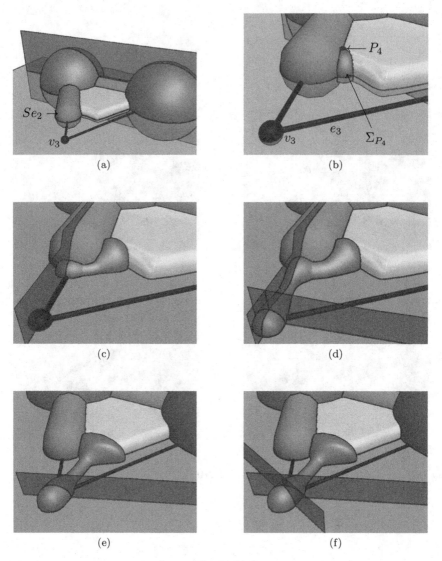

Fig. B.11

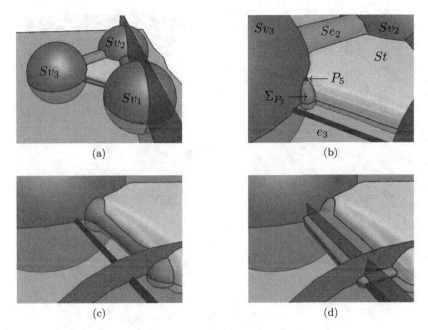

Fig. B.12

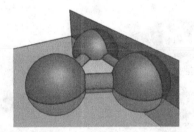

Fig. B.13

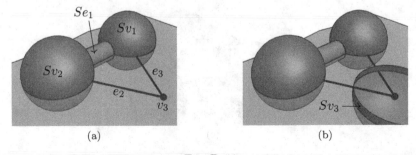

Fig. B.14

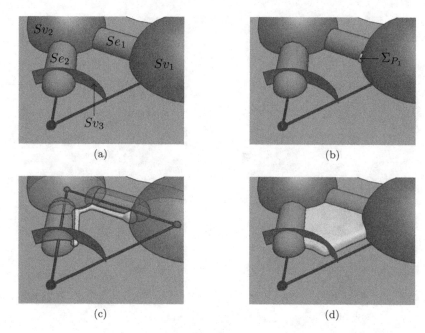

Fig. B.15

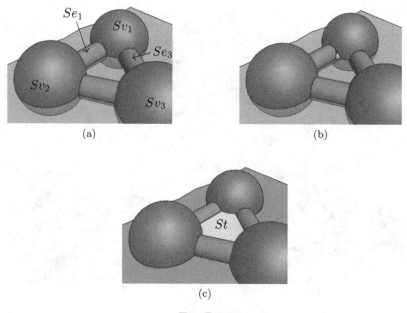

Fig. B.16

Appendix C

Proof of Proposition 6.57

Let $g : S_2 \to M$ be a parametrization of Σ_2 and let T be a good triangulation of M with respect to f and g. Then, T shells each region of Σ_1 and each region of Σ_2. By Proposition 6.41, the union $\Sigma_1 \cup \Sigma_2 \cup \Sigma_T$ is a regular filling Dehn surface of M.

Take an inflation f' of T from f, and define $\Sigma_1' = f'(S_1)$. Making the spiral pipings involved in the transformation of $\Sigma_1 \cup \Sigma_T$ into Σ_1' small enough it is possible to asume that they do not meet Σ_2.

Since regularity is preserved by spiral piping (Proposition 2.16), the union $\Sigma_1' \cup \Sigma_2$ is a regular filling Dehn surface of M.

Let R_1 be a region of Σ_1' such that $R_1 \cap \Sigma_2 \neq \emptyset$, and assume first that there are no spiral pipings around triple points of ∂R_1.

Let $\widetilde{R}_1$ be the region of Σ_T containing R_1. By Sec. 6.4.3, the cellular decomposition induced by $\Sigma_1 \cup \Sigma_2$ on $\mathrm{cl}(\widetilde{R}_1)$ is similar to that induced by one, two or three coordinate planes on the closed unit 3-disk of euclidean space. Using this model it is clear that Σ_2 induces a shellable cellular decomposition of $\mathrm{cl}(R_1)$.

Figure 2.9 shows that the effect of a spiral piping around a triple point P of $\Sigma_1 \cup \Sigma_T$ in a region R_1 incident with P is the addition or subtraction of a small 3-disk in a neighbourhood of P. Moreover, this operation do not destroys the shellability of the cell decomposition induced in R_1 by Σ_2. Then, if several of these spiral pipings around triple points of $\Sigma_1 \cup \Sigma_T$ belonging to ∂R_1 are introduced, and R_1' is the region resulting from R_1 after these spiral pipings, the Dehn surface Σ_2 still induces a shellable cellular decomposition of $\mathrm{cl}(R_1')$. This proves that Σ_2 shells Σ_1'.

The most tedious part is to check that Σ_1' shells each region of Σ_2. Let R_2 be a region of Σ_2, and define $B_2 = \mathrm{cl}(R_2)$. Since T is a good triangulation with respect to f and g, it shells R_2 and the triangulation T_{B_2}

251

induced by T in B_2 is collapsable.

Consider first B_2 endowed with the cell decomposition Γ induced by $\Sigma_1 \cup \Sigma_T$ without spiral pipings. We will define a shelling of Γ following a collapsing of T_{B_2} to a point.

Let

$$B_2 \searrow \overset{s_3}{\cdots} \searrow B_2^2 \searrow \overset{s_2}{\cdots} \searrow B_2^1 \searrow \overset{s_1}{\cdots} \searrow w \tag{1}$$

be a collapsing of T_{B_2} to a point. As in the proof of Key Lemma 2, it is possible to assume that the first s_3 collapsings of the sequence (1) are 3-collapsings of B_2 and the resulting complex B_2^2 is a simplicial 2-complex, the next s_2 collapsings of (1) are 2-collapsings of B_2^2, and the resulting complex B_2^1 is a simplicial 1-complex, and finally the last s_1 collapsings of (1) are 1-collapsings of B_2^1 and the result is a single vertex $P \in T_{B_2}$.

For each collapsing $(\epsilon^i, \epsilon^{i-1})$ of (1), a sequence $\text{sh}(\epsilon^i, \epsilon^{i-1})$ of 3-cells of B_2 will be defined. The concatenation of them will provide a shelling of Γ.

Case 1: 3-collapsings. If (σ, t) is a 3-collapsing of (1) (Figs. C.1(a) and C.1(b)), there are three possibilities:

(1) If $t \subset \Sigma_2$, then Σ_2 splits $R(t)$ into two 6-gonal prisms E_1 and E_2, and only one of them, E_1 for example, is contained in B_2. In this case, define $\text{sh}(\sigma, t) = (E_1, R(\sigma))$.

(2) If $t \subset \text{int}(B_2)$ and $t \not\subset \Sigma_1$, then define $\text{sh}(\sigma, t) = (R(t), R(\sigma))$ (see Figs. C.1(c), C.1(d) and C.2).

(3) If $t \subset \text{int}(B_2)$ and $t \subset \Sigma_1$, then Σ_1 divides $R(t)$ into two 6-gonal prisms F_1 and F_2, and only one of them, say F_1, is contained in the tetrahedron σ. In this case define $\text{sh}(\sigma, t) = (F_2, F_1, R(\sigma))$.

Case 2: 2-collapsings. Let (τ, e) be a 2-collapsing of (1) (see Figs. C.3(a) and C.3(b)). There are two subcases.

Case 2.1: $e \subset \text{int}(B_2)$. In this case $\tau \subset \text{int}(B_2)$.

Case 2.1.1: $e \not\subset \Sigma_1$. Therefore $\tau \not\subset \Sigma_1$. Let $q_1, \ldots, q_n$ be the triangles of T incident with e and distinct from τ. Define

$$\text{sh}(\tau, e) = \big(R(e, q_1), R(e, q_2), \ldots, R(e, q_n), R(e), R(e, \tau), R(\tau)\big),$$

where the order of $q_1, \ldots, q_n$ is unimportant (Figs. C.3(c) to C.3(f)).

Figure C.4(a) shows how the intersection of T and Σ_T with a plane "orthogonal" to e through the midpoint of e looks like. This schematic view will be used all along Case 2.

Case 2.1.2: $e \subset \Sigma_1$. The sequence $\mathrm{sh}(\tau, e)$ should be modified in the following way. Since $e \subset \Sigma_1$, two or four among the triangles $q_1, \ldots, q_n$ are contained in Σ_1 depending on whether $e \not\subset \mathrm{Sing}(\Sigma_1)$ or $e \subset \mathrm{Sing}(\Sigma_1)$, respectively.

If $q_i \subset \Sigma_1$ for some $i = 1, \ldots, n$, the surface Σ_1 divides the 2-gonal prism $R(e, q_i)$ into two 3-gonal prisms E_i^1 and E_i^2 (Fig. C.4(b)). The pair of regions E_i^1, E_i^2 replace the region $R(e, q_i)$ in the sequence $\mathrm{sh}(\tau, e)$. Repeat the same process for each $i \in \{1, \ldots, n\}$ such that $q_i \subset \Sigma_1$.

Furthermore, the Dehn surface Σ_1 divides the region $R(e)$ in either two regions F_1 and F_2 of $\Sigma_1 \cup \Sigma_2 \cup \Sigma_T$, if $e \not\subset \mathrm{Sing}(\Sigma_1)$, or four regions F_1, F_2, F_3 and F_4 of $\Sigma_1 \cup \Sigma_2 \cup \Sigma_T$, if $e \subset \mathrm{Sing}(\Sigma_1)$. If $e \not\subset \mathrm{Sing}(\Sigma_1)$) other two possibilities arise:

(i) If $\tau \not\subset \Sigma_1$ (Fig. C.5(a)), only one of the two regions F_1 or F_2 has a face contained in $S\tau$. Assume that F_2 is that region.
(ii) If $\tau \subset \Sigma_1$ (Fig. C.5(b)), then both F_1 and F_2 have an edge face in $S\tau$, and therefore the order of F_1 and F_2 is irrelevant.

Replace the region $R(e)$ with the pair of regions F_1, F_2 in $\mathrm{sh}(\tau, e)$.

If $e \subset \mathrm{Sing}(\Sigma_1)$, assume without loss of generality that the regions F_1, F_2, F_3 and F_4 are cyclically ordered around e (Figs. C.6(a) and C.6(b)). In this case, there is a slight difference between the situations $\tau \not\subset \Sigma_1$ and $\tau \subset \Sigma_1$:

(i) If $\tau \not\subset \Sigma_1$ (Fig. C.6(a)), only one of the four regions F_1, F_2, F_3, F_4 has an face in $S(\tau)$. Assume that F_4 is that region.
(ii) If $\tau \subset \Sigma_1$ (Fig. C.6(b)), then two of the regions F_1, F_2, F_3, F_4 has an face in $S(\tau)$. Assume that F_4 to be one of them.

Replace $R(e)$ with the sequence of four regions F_1, F_2, F_3, F_4 in $\mathrm{sh}(\tau, e)$.

If $\tau \subset \Sigma_1$, the triangle τ (and therefore Σ_1) splits $R(e, \tau)$ into two regions G_1 and G_2 of $\Sigma_1 \cup \Sigma_2 \cup \Sigma_T$. It also divides $R(\tau)$ into two regions H_1 and H_2 of $\Sigma_1 \cup \Sigma_2 \cup \Sigma_T$ (Figs. C.5(b) and C.6(b)). The regions $R(e, \tau)$ and $R(\tau)$ should be replaced by G_1, G_2 and H_1, H_2, respectively, in $\mathrm{sh}(\tau, e)$.

Case 2.2: $e \subset \partial B_2$. The proof of this case is divided into several sub-cases summarized in the following scheme:

$$
\begin{cases}
e \not\subset \mathrm{Sing}(\Sigma_2) \begin{cases}
e \not\subset \Sigma_1 \begin{cases} \tau \not\subset \Sigma_2 \\ \tau \subset \Sigma_2 \end{cases} \\[1em]
e \subset \Sigma_1 \begin{cases} \tau \not\subset \Sigma_1 \cup \Sigma_2 \\ \tau \subset \Sigma_1 \\ \tau \subset \Sigma_2 \end{cases}
\end{cases} \\[3em]
e \subset \mathrm{Sing}(\Sigma_2) \qquad\quad \begin{cases} \tau \not\subset \Sigma_2 \\ \tau \subset \Sigma_2 \end{cases}
\end{cases}
$$

The possible configurations of $\Sigma_1 \cup \Sigma_2 \cup \Sigma_T$ around e and τ are similar to those in Case 2.1, but now the regions not contained in B_2 are ignored (Fig. C.7). It is possible to construct the sequence $\mathrm{sh}(\tau, e)$ in the following way: consider for a moment that $\Sigma_2 \subset \Sigma_1$, and construct a preliminary sequence of 3-cells $\widetilde{\mathrm{sh}(\tau, e)}$ following the rules of the Case 2.1. Then, remove from $\widetilde{\mathrm{sh}(\tau, e)}$ the regions not contained in B_2. The resulting sequence is $\mathrm{sh}(\tau, e)$.

Case 3: 1-collapsings. Let (ε, v) be a 1-collapsing of the sequence (**1**). Assume first that $v \notin \Sigma_1 \cup \Sigma_2$. This implies that v is an interior vertex of B_2. Let $a_1, \ldots, a_n$ be all the edges of T different from ε and incident with v. Let $\omega_1, \ldots, \omega_\ell$ be all the triangles of T incident with v, and assume that the first $k < \ell$ of them are the triangles of T incident with ε. Let $r_1, \ldots, r_m$ be all the regions of the form $R(v, a_i, \omega_j)$.

Consider

$$
\mathrm{sh}(\varepsilon, v) = R(v, a_1), \ldots, R(v, a_n), r_1, r_2, \ldots
$$
$$
\ldots, r_m, R(v, \omega_1), \ldots, R(v, \omega_\ell), R(v), R(v, \varepsilon, \omega_1), R(v, \varepsilon, \omega_2), \ldots
$$
$$
\ldots, R(v, \varepsilon, \omega_k), R(v, \varepsilon), R(\varepsilon, \omega_1), \ldots, R(\varepsilon, \omega_\ell), R(\varepsilon),
$$

see Figs. C.8 to C.10; in Fig. C.9, v' denotes the other vertex of ε different from v.

Consider now the case $v \in \Sigma_1$.

Let $i = 1, \ldots, n$ be such that $a_i \subset \Sigma_1$. If $a_i \not\subset \mathrm{Sing}(\Sigma_1)$, the Dehn surface Σ_1 divides $R(v, a_i)$ into two regions E_1 and E_2 of $\Sigma_1 \cup \Sigma_2 \cup \Sigma_T$. Replace the region $R(v, a_i)$ with the pair E_1, E_2 in $\mathrm{sh}(\varepsilon, v)$. If $a_i \subset \mathrm{Sing}(\Sigma_1)$, then Σ_1 divides $R(v, a_i)$ into four regions E_1, E_2, E_3 and E_4. In this case, order

$E_1, \ldots, E_4$ cyclically around a_i and replace $R(v, a_i)$ with E_1, E_2, E_3, E_4 in $\mathrm{sh}(\varepsilon, v)$.

Let ω_j (with $j = 1, \ldots, \ell$) be a triangle of T incident with v and such that $\omega_j \subset \Sigma_1$. If R is a region of type $R(v, \cdot, \omega_j)$, $R(v, \omega_j)$ or $R(\varepsilon, \omega_j)$ (if $\varepsilon < \omega_j$), then Σ_1 divides R into two regions F_1 and F_2 of $\Sigma_1 \cup \Sigma_2 \cup \Sigma_T$. The region R is replaced in $\mathrm{sh}(\varepsilon, v)$ with the pair or regions F_1, F_2.

Since $v \in \Sigma_1$, the Dehn surface Σ_1 divides $R(v)$ into the regions $G_1, \ldots, G_h$ of $\Sigma_1 \cup \Sigma_2 \cup \Sigma_T$, where $h = 2$, 4 or 8 depending on whether v is a simple, double or triple point of Σ_1. In any case, $G_1, \ldots, G_h$ are the 3-cells of a shellable cellular decomposition of $R(v)$. Assume that the ordering $G_1, \ldots, G_h$ is a shelling of $R(v)$ and that the last region G_h has a face on the edge sphere $S\varepsilon$. In this conditions, replace $R(v)$ with $G_1, \ldots, G_h$ in $\mathrm{sh}(\varepsilon, v)$.

If $\varepsilon \subset \Sigma_1$ and $R = R(v, \varepsilon)$ or $R = R(\varepsilon)$, then Σ_1 divides R into two or four regions of $\Sigma_1 \cup \Sigma_2 \cup \Sigma_T$ depending on whether $\varepsilon \not\subset \mathrm{Sing}(\Sigma_1)$ or $\varepsilon \subset \mathrm{Sing}(\Sigma_1)$. Proceed as with the regions of type $R(v, a_i)$. If Σ_1 divides R into two regions H_1 and H_2 of $\Sigma_1 \cup \Sigma_2 \cup \Sigma_T$, replace R with H_1, H_2 in the sequence in $\mathrm{sh}(\varepsilon, v)$. On the other hand, if Σ_1 divides R into four regions H_1, H_2, H_3 and H_4 of $\Sigma_1 \cup \Sigma_2 \cup \Sigma_T$, assume that they are cyclically ordered around ε, and replace R with H_1, H_2, H_3, H_4 in $\mathrm{sh}(\varepsilon, v)$.

If v belongs to Σ_2, three possibilities arise depending on whether v is a simple (Fig. C.11(a)), double (Fig. C.11(b)) or triple point of Σ_2 (Fig. C.11(c)). Furthermore, if v is a simple or double point of Σ_1 we must distinguish between the different relative positions of v and ε with respect to Σ_1. Proceed in a similar way as in the case $v \in \mathrm{int}(B_2)$, but disregarding the regions of $\Sigma_1 \cup \Sigma_2 \cup \Sigma_T$ outside B_2. As in Case 2.2, assume for a moment that both v and ϵ are inside of B_2 and $\Sigma_2 \subset \Sigma_1$. With these assumptions, it is possible to build the sequence $\widetilde{\mathrm{sh}}(\varepsilon, v)$ with the regions of $\Sigma_1 \cup \Sigma_2 \cup \Sigma_T$ contained in $(Bv \cup B\varepsilon) - Bv'$ as in the case $v \in \mathrm{int}(B_2)$. Once this is done, $\mathrm{sh}(\varepsilon, v)$ is obtained by removing from $\widetilde{\mathrm{sh}}(\varepsilon, v)$ all the regions outside B_2 (see Figs. C.12)

Case 4: the last vertex P. Finally, consider the remaining vertex w after collapsing B_2 using the sequence (1). The structure of $\mathrm{sh}(w)$ is quite similar to the one of the sequences $\mathrm{sh}(\varepsilon, v)$ of Case 3, with the exception that $\mathrm{sh}(w)$ only contains the regions of $\Sigma_1 \cup \Sigma_2 \cup \Sigma_T$ contained in Bw.

For simplicity, assume that $w \in \mathrm{int}(B_2)$ and $w \notin \Sigma_1$.

Denote by:

- $a_1, \ldots, a_n$ the edges of T incident with w.
- $\omega_1, \ldots, \omega_\ell$ all the triangles of T incidents with w.
- $r_1, \ldots, r_m$ all the regions of Σ_T of type $R(w, \cdot, \cdot)$.

Then,

$$\text{sh}(w) = R(w, a_1), \ldots, R(w, a_n), r_1, \ldots,$$
$$\ldots r_m, R(P, \omega_1), \ldots, R(P, \omega_l), R(P).$$

If $w \in \Sigma_1$, as in Cases 1 to 3, each region R of $\text{sh}(w)$ intersecting Σ_1 should be replaced in $\text{sh}(w)$ in the proper way with the collection of regions of $\Sigma_1 \cup \Sigma_2 \cup \Sigma_T$ contained in R. If $w \in \Sigma_2$, build a temporary sequence $\widetilde{\text{sh}(w)}$ as if $P \in B_2$ and $\Sigma_2 \subset \Sigma_1$, and after that construct $\text{sh}(w)$ by removing from $\widetilde{\text{sh}(w)}$ the regions not contained in B_2.

Let

$$(\sigma_1, t_1), \ldots, (\sigma_{s_3}, t_{s_3}), \quad (\tau_1, e_1), \ldots, (\tau_{s_2}, e_{s_2}), \quad (\varepsilon_1, v_1), \ldots, (\varepsilon_{s_1}, v_{s_1})$$

be the 3-collapsings, 2-collapsings and 1-collapsings of (1), respectively. It is not difficult to check that

$$\text{sh}(\Gamma) = \text{sh}(\sigma_1, t_1) * \cdots * \text{sh}(\sigma_{s_3}, t_{s_3}) *$$
$$\text{sh}(\tau_1, e_1) * \cdots * \text{sh}(\tau_{s_2}, e_{s_2}) *$$
$$\text{sh}(\varepsilon_1, v_1) * \cdots * \text{sh}(\varepsilon_{s_1}, v_{s_1}) * \text{sh}(w)$$

is a shelling of the cellular decomposition Γ of $B_2 = \text{cl}(R_2)$ induced by $\Sigma_1 \cup \Sigma_T$.

It is also easy to see that the chosen sequence $\text{sh}(\Gamma)$ is not unique. In particular, in Case 3, it is possible to reorder in any way the regions $R(v, a_1), \ldots, R(v, a_n), r_1, \ldots, r_m$ at the beginning of the sequence $\text{sh}(\varepsilon, v)$. The same happens with the regions $R(v, \varepsilon, \omega_1), R(v, \varepsilon, \omega_2), \ldots, R(v, \varepsilon, \omega_k), R(v, \varepsilon)$ in the middle part of $\text{sh}(\varepsilon, v)$. In the same way, in Case 4 it is possible to reorder the regions $R(w, a_1), \ldots, R(w, a_n), r_1, \ldots, r_m$ at the beginning of $\text{sh}(P)$.

Consider now the cell decomposition Γ' of B_2 induced by Σ_1'. Let $Q \in R_2$ be a triple point of $\Sigma_1 \cup \Sigma_T$ around which there is a spiral piping of f'. By the regularity of $\Sigma_1 \cup \Sigma_T$, there are eight different regions $W_1, \ldots, W_8$ of $\Sigma_1 \cup \Sigma_2 \cup \Sigma_T$, incident with Q. Assume that $W_1, \ldots, W_8$ are ordered as they appear in the shelling $\text{sh}(\Gamma)$ constructed above. Let $W_1', \ldots, W_8'$ be the regions of $\Sigma_1' \cup \Sigma_2$ that coincide with $W_1, \ldots, W_8$, respectively, outside the spiral pipings of Σ_1'. In six cases, W_i' is obtained around Q after removing from W_i a 2-gonal prism $\widetilde{W}_i$ enclosed by $\Sigma_1 \cup \Sigma_T$ and the piping. In the remaining two cases W_i' is obtained around Q attaching to W_i two of those 2-gonal prisms. Besides $W_1', \ldots, W_8'$, there exist i_1 and $i_2 \in \{1, \ldots, 8\}$ such that the 2-gonal prisms $\widetilde{W}_{i_1}$ and $\widetilde{W}_{i_2}$ become regions of Σ_1'. Call these two regions the *piping regions* of the spiral piping around Q, and assume that $i_1 < i_2$.

Firstly, modify $\mathrm{sh}(\Gamma)$ as follows:

- for $i \neq i_1, i_2$, W_i is replaced with W_i';
- W_{i_1} is replaced with $W_{i_1}', \widetilde{W}_{i_1}$; and
- W_{i_2} is replaced with $\widetilde{W}_{i_2}, W_{i_2}'$.

This operation is the *adaptation of* $\mathrm{sh}(\Gamma)$ around Q. Adapt $\mathrm{sh}(\Gamma)$ around every triple point of $\Sigma_1 \cup \Sigma_T$ in R_2 that becomes affected by the spiral pipings, obtaining a sequence $\mathrm{sh}(\Gamma')$ that contains all the 3-cells of Γ'.

Let B_Q be a closed 3-disk standardly embedded in M centred at Q such that B_Q intersects each of the sheets of $\Sigma_1 \cup \Sigma_T$ through Q in a closed 2-disk, and such that the spiral piping around Q lies inside B_Q. Let Γ_Q' be the cell decomposition induced by Σ_1' in B_Q, and let sh_Q be the ordering of the 3-cells of Γ_Q' induced by $\mathrm{sh}(\Gamma')$ in the obvious way. The sequence $\mathrm{sh}(\Gamma')$ is *good* at Q if sh_Q is a shelling of Γ_Q', otherwise $\mathrm{sh}(\Gamma)$ is *bad* at Q. It is clear that $\mathrm{sh}(\Gamma')$ is a shelling of Γ' if and only if it is good at every triple point of $\Sigma_1 \cup \Sigma_T$ in R_2 with a spiral piping around it.

Assume that $\mathrm{sh}(\Gamma')$ is bad at Q. Set $\Sigma_Q = \partial B_Q$, and consider the filling Dehn surface $\Sigma_1' \# \Sigma_Q$ obtained after connecting Σ_1' and Σ_Q around a double point Q' of Σ_1' incident with W_8 as in Fig. C.13(b). Inflating Q' from Σ_1' (Fig. C.13(a)) and applying a piping passing move, $\Sigma_1' \# \Sigma_Q$ is obtained. Let $j_1, j_2 \in \{1, \ldots 7\}$, with $j_1 < j_2$, be such that the two piping regions $\widetilde{W}_{j_1}'$ and $\widetilde{W}_{j_2}'$ of the spiral piping around Q' are contained in the regions W_{j_1}' and W_{j_2}' of $\Sigma_1' \cup \Sigma_2$, respectively. For $i = 1, \ldots, 8$, the 2-sphere Σ_Q splits W_i' into two regions of $(\Sigma_1' \# \Sigma_Q) \cup \Sigma_2$: one outside B_Q, again denoted by W_i', and the other inside B_Q, denoted by W_i''. Modify $\mathrm{sh}(\Gamma')$ as follows:

- W_j' remains unchanged for $j \in \{1, \ldots, 7\}, j \neq j_1, j_2$;
- $\widetilde{W}_{i_1}$ and $\widetilde{W}_{i_2}$ are removed from $\mathrm{sh}(\Gamma')$
- W_{j_1}' is replaced with the pair $W_{j_1}', \widetilde{W}_{j_1}'$;　　·
- W_{j_2}' is replaced with the pair $\widetilde{W}_{j_2}', W_{j_2}'$; and finally
- W_8' with sh_Q, W_8';

where sh_Q is an ordering of the regions of $\Sigma_1' \# \Sigma_Q$ contained in B_Q as the one given by the following Claim:

Claim C.34. *There is a shelling* sh_Q *of* Γ_Q' *that finishes at* W_8''.

Proof. Consider W_{i_1}'', W_{i_2}'', $\widetilde{W}_{i_1}$ and $\widetilde{W}_{i_2}$ as above. Benote by $W_{i_3}'', \ldots W_{i_8}''$ the remaining 3-cells of Γ_Q' as it is schematically depicted in Fig. C.15(a). Up to symmetry, it is clear that the regions W_{i_1}'' and W_{i_2}'' are equivalent. In the same way, W_{i_4}'', W_{i_7}'' and W_{i_8}'' are equivalent to W_{i_3}'', and W_{i_6}'' is equivalent to W_{i_5}''. If $W_8'' = W_{i_1}''$, take (Fig. C.16)

$$\mathrm{sh}_Q = (W_{i_3}'', W_{i_5}'', W_{i_6}'', W_{i_4}'', W_{i_8}'', W_{i_2}'', \widetilde{W}_{i_2}, \widetilde{W}_{i_1}, W_{i_7}'', W_{i_1}'').$$

If $W_8'' = W_{i_7}''$ it suffices with swapping W_{i_1}'' and W_{i_7}'' in sh_Q above. If $W_8'' = W_{i_5}''$ take

$$\mathrm{sh}_Q = (W_{i_1}'', \widetilde{W}_{i_1}, \widetilde{W}_{i_2}, W_{i_2}'', W_{i_3}'', W_{i_4}'', W_{i_7}'', W_{i_8}'', W_{i_6}'', W_{i_5}'').$$

Obviously, the election of the the sequences sh_Q is not unique. $\square$

Making the same modification for all triple point of $\Sigma_1 \cup \Sigma_T$ in R_2 with a spiral piping around it at which $\mathrm{sh}(\Gamma')$ is bad, a new filling Dehn surface that shells R_2 is obtained. Repeating the same operation for all regions of Σ_2, we get a filling Dehn surface Σ_1'' $\boldsymbol{f}$-homotopic to Σ_1 that shells Σ_2. Since the modifications that transform Σ_1' into Σ_1'' are done around triple points of $\Sigma_1 \cup \Sigma_T$, it is clear that Σ_2 still shells Σ_1''. This finishes the proof. $\square$

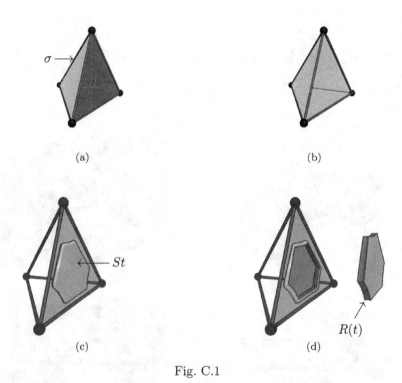

(a)

(b)

(c)

(d)

Fig. C.1

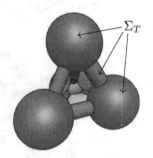

Fig. C.2

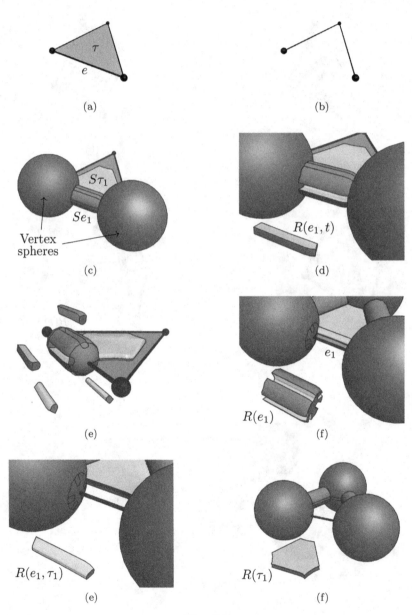

Fig. C.3

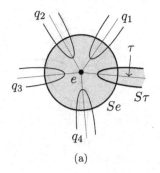

(a)

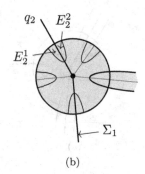

(b)

Fig. C.4

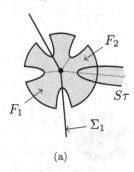

(a)

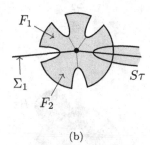

(b)

Fig. C.5

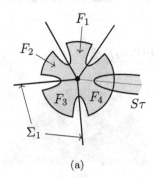

(a)

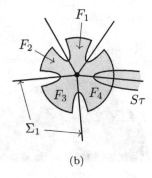

(b)

Fig. C.6

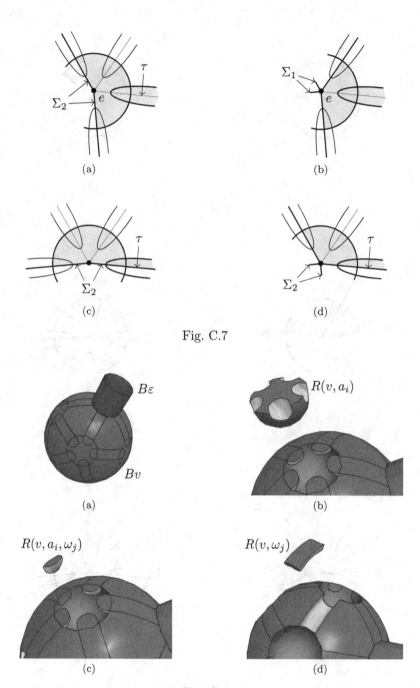

Fig. C.7

Fig. C.8

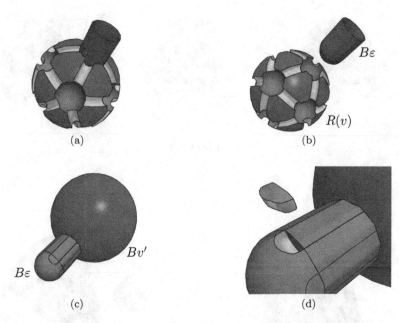

Fig. C.9

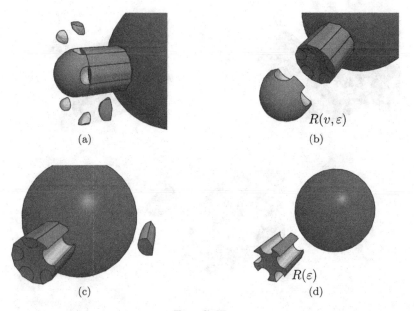

Fig. C.10

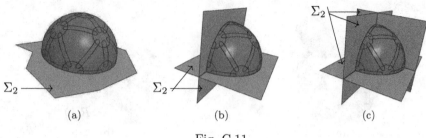

(a)   (b)   (c)

Fig. C.11

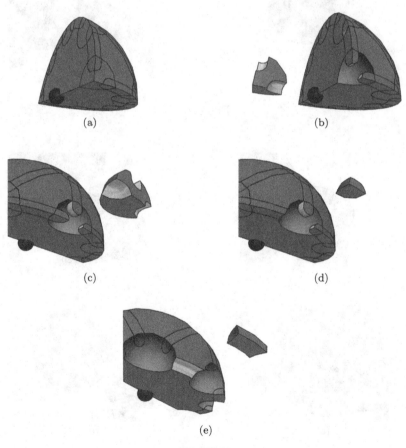

(a)   (b)

(c)   (d)

(e)

Fig. C.12

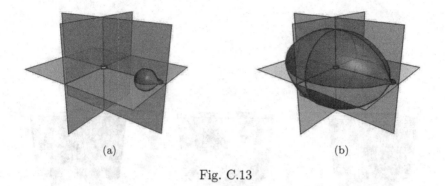

(a) (b)

Fig. C.13

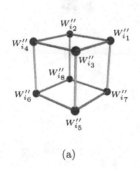

(a)

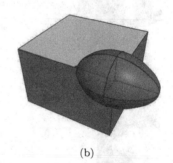

(b)

Fig. C.14

(a)

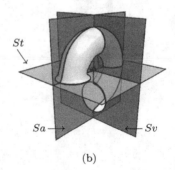

(b)

Fig. C.15

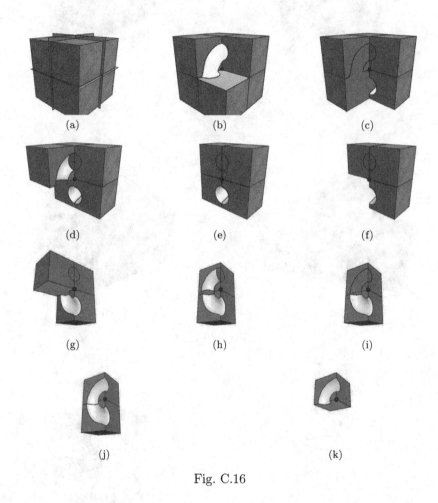

(a)   (b)   (c)

(d)   (e)   (f)

(g)   (h)   (i)

(j)   (k)

Fig. C.16

Bibliography

[1] I. Agol, The virtual Haken conjecture, *Documenta Mathematica* **18**, pp. 1045–1087 (2013).

[2] I. R. Aitchison, S. Matsumoto and J. H. Rubinstein, Immersed surfaces in cubed manifolds, *Asian Journal of Mathematics* **1**, 1, pp. 85–95 (1997).

[3] G. Amendola, A local calculus for nullhomotopic filling Dehn spheres. *Algebraic and Geometric Topology* **9**, 2, pp. 903–933 (2009).

[4] G. Amendola, A 3-manifold complexity via immersed surfaces, *Journal of Knot Theory and its Ramifications* **19**, 12, pp. 1549–1569 (2010).

[5] G. Amendola, Orientable closed 3-manifolds with surface-complexity one, *Atti del Seminario Matematico e Fisico dell' Università di Modena e Reggio Emilia* **57**, pp. 17–26 (2010).

[6] E. K. Babson and C. S. Chan, Counting faces of cubical spheres modulo two, *Discrete Mathematics* **212**, 3, pp. 169–183 (2000).

[7] T. F. Banchoff, Triple points and surgery of immersed surfaces, *Proceedings of the American Mathematical Society* **46**, 3, pp. 407–413 (1974).

[8] R. H. Bing, Some aspects of the topology of 3-manifolds related to the Poincaré conjecture, in *Lectures on modern mathematics, Vol. II*. Wiley, pp. 93–128 (1964).

[9] H. Bruggesser and P. Mani, Shellable decompositions of cells and spheres, *Math. Scand.* **29**, pp. 197–205 (1971).

[10] J. S. Carter, Extending immersions of curves to properly immersed surfaces, *Topology and its Applications* **40**, pp. 287–306 (1991).

[11] J. S. Carter and M. Saito, *Knotted surfaces and their diagrams*, Mathematical Surveys and Monographs, Vol. 55. American Mathematical Society (1998).

[12] J. S. Carter and M. Saito, Surfaces in 3-space that do not lift to embeddings in 4-space, in *Knot theory (Warsaw, 1995), Banach Center Publications*, Vol. 42, pp. 29–47 (1998).

[13] J. Cerf, Topologie de certains espaces de plongements, *Bulletin de la Société Mathématique de France* **89**, pp. 227–380 (1961).

[14] D. R. J. Chillingworth, Collapsing three-dimensional convex polyhedra, *Mathematical Proceedings of the Cambridge Philosophical Society* **63**, pp. 353–357 (1967).

[15] R. D. Edwards and R. C. Kirby, Deformations of spaces of embeddings, *Annals of Mathematics* **2**, 93, pp. 63–88 (1971).

[16] R. Fenn and C. Rourke, Nice spines of 3-manifolds, in R. Fenn (ed.), *Topology of Low-Dimensional Manifolds, Lecture Notes in Mathematics*, Vol. 722. Springer, pp. 31–36 (1979).

[17] L. Funar, Cubulations, immersions, mappability and a problem of Habegger, *Annales Scientifiques de l'École Normale Supérieure* **32**, 5, pp. 681–700 (1999).

[18] C. A. Giller, Towards a classical knot theory for surfaces in $\mathbb{R}^4$, *Illinois Journal of Mathematics* **26**, 4, pp. 591–631 (1982).

[19] L. Glaser, *Geometrical combinatorial topology*, no. 1 in Geometrical Combinatorial Topology. Van Nostrand Reinhold (1970).

[20] M. Golubitsky and V. Guillemin, *Stable mappings and their singularities, Graduate Texts in Mathematics*, Vol. 14. Springer (1973).

[21] S. Goodman and M. Kossowski, Immersions of the projective plane with one triple point, *Differential Geometry and its Applications* **27**, 4, pp. 527–542 (2009).

[22] V. Guillemin and A. Pollack, *Differential topology.* Prentice-Hall (1974).

[23] W. Haken, On homotopy 3-spheres, *Illinois Journal of Mathematics* **10**, pp. 159–180 (1966).

[24] W. Haken, Some special presentations of homotopy 3-spheres, in *Topology Conference (Virginia Polytech. Inst. and State Univ., Blacksburg, Va., 1973).* Springer, pp. 97–107. Lecture Notes in Math., Vol. 375 (1974).

[25] J. M. Hall, *The theory of groups.* Chelsea (1976), reprinting of the 1968 edition.

[26] J. Hass and J. Hughes, Immersions of surfaces in 3-manifolds, *Topology* **24**, 1, pp. 97–112 (1985).

[27] J. Hempel, *3-Manifolds, Annals of Mathematics Studies*, Vol. 86. Princeton University Press; University of Tokyo Press (1976).

[28] H. M. Hilden, M. T. Lozano and J. M. Montesino, Universal knots, *Bulletin of the American Mathematical Society* **8**, 3, pp. 449–450 (1983).

[29] H. M. Hilden, M. T. Lozano and J. M. Montesinos, On knots that are universal, *Topology* **24**, 4, pp. 499–504 (1985).

[30] M. W. Hirsch, *Differential topology, Graduate Texts in Mathematics*, Vol. 33. Springer (1976).

[31] T. Homma, Elementary deformations on homotopy spheres, in *Topology and computer science (Atami, 1986).* Kinokuniya, pp. 21–27 (1987).

[32] T. Homma and T. Nagase, On elementary deformations of maps of surfaces into 3-manifolds. I, *Yokohama Math. J.* **33**, 1-2, pp. 103–119 (1985).

[33] T. Homma and T. Nagase, On elementary deformations of maps of surfaces into 3-manifolds. II, in *Topology and computer science (Atami, 1986).* Kinokuniya, pp. 1–20 (1987).

[34] J. F. P. Hudson, *Piecewise linear topology*, University of Chicago Lecture Notes prepared with the assistance of J. L. Shaneson and J. Lees. W. A. Benjamin (1969).

[35] J. F. P. Hudson and E. C. Zeeman, On combinatorial isotopy, *Institut des Hautes tudes Scientifiques. Publications Mathmatiques* **19**, pp. 69–94 (1964).

[36] I. Johansson, Über singuläre elementarflächen und das Dehnsche lemma. *Mathematische Annalen* **110**, pp. 312–320 (1934).

[37] I. Johansson, Über singuläre elementarflächen und das Dehnsche lemma. II. *Mathematische Annalen* **115**, pp. 658–669 (1938).

[38] A. Kazarov, 3-*manifolds of cubic complexity* 2, Master's thesis.

[39] R. Kirby, Problems in low-dimensional topology, in *Geometric topology (Athens, GA, 1993), AMS/IP Stud. Adv. Math.*, Vol. 2. American Mathematical Society, pp. 35–473 (1997).

[40] A. A. Kosinski, *Differential manifolds, Pure and Applied Mathematics*, Vol. 138. Academic Press (1993).

[41] B. H. Li, On immersions of manifolds in manifolds, *Scientia Sinica. Series A. Mathematical, Physical, Astronomical & Technical Sciences* **25**, 3, pp. 255–263 (1982).

[42] Á. Lozano Rojo and R. Vigara Benito, The triple point spectrum of closed orientable 3-manifolds, (2014), arXiv:1412.1637.

[43] Á. Lozano Rojo and R. Vigara Benito, On the subadditivity of montesinos complexity of closed orientable 3-manifolds, *Revista de la Real Academia de Ciencias Exactas, Físicas y Naturales. Serie A. Matemáticas* **109**, 2, pp. 267–279 (2015).

[44] Á. Lozano Rojo and R. Vigara Benito, Representing knots by filling dehn spheres, *Journal of Knot Theory and its Ramifications* (to appear 2016), arXiv:1508.06295.

[45] W. S. Massey, *A basic course in algebraic topology, Graduate Texts in Mathematics*, Vol. 127. Springer (1991).

[46] J. N. Mather, Stability of C^∞ mappings. II. Infinitesimal stability implies stability, *Annals of Mathematics* **89**, pp. 254–291 (1969).

[47] J. Milnor, A unique decomposition theorem for 3-manifolds, *American Journal of Mathematics* **84**, pp. 1–7 (1962).

[48] E. E. Moise, Affine structures in 3-manifolds. V. The triangulation theorem and Hauptvermutung, *Annals of Mathematics* **56**, pp. 96–114 (1952).

[49] J. M. Montesinos-Amilibia, Representing 3-manifolds by Dehn spheres, in *Mathematical contributions: volume in honor of Professor Joaquín Arregui Fernández (Spanish)*, Homen. Univ. Complut. Editorial Complutense, pp. 239–247 (2000).

[50] J. R. Munkres, *Elementary differential topology, Lectures given at Massachusetts Institute of Technology, Fall*, Vol. 1961. Princeton University Press (1966).

[51] C. D. Papakyriakopoulos, On Dehn's lemma and the asphericity of knots, *Annals of Mathematics* **66**, pp. 1–26 (1957).

[52] G. Perelman, The entropy formula for the ricci flow and its geometric applications, (2002), arXiv:math/0211159.

[53] G. Perelman, Finite extinction time for the solutions to the Ricci flow on certain three-manifolds, (2003b), arXiv:math/0307245.

[54] G. Perelman, Ricci flow with surgery on three-manifolds, (2003a), arXiv:math/0303109.

[55] T. Radó, Über den begriff der riemannsche fläche, *Acta Scientarum Mathematicarum Universitatis Szegediensis* **2**, pp. 101–121 (1924).

[56] D. Rolfsen, *Knots and links, Mathematics Lecture Series*, Vol. 7. Publish or Perish (1976).

[57] D. Roseman, Reidemeister-type moves for surfaces in four-dimensional space, in *Knot theory (Warsaw, 1995), Banach Center Publ.*, Vol. 42. Polish Academy of Sciences, pp. 347–380 (1998).

[58] C. P. Rourke and B. J. Sanderson, *Introduction to piecewise-linear topology*, Springer Study Edition. Springer (1982).

[59] M. E. Rudin, An unshellable triangulation of a tetrahedron, *Bulletin of the American Mathematical Society* **64**, pp. 90–91 (1958).

[60] D. E. Sanderson, Isotopy in 3-manifolds. I. Isotopic deformations of 2-cells and 3-cells, *Proceedings of the American Mathematical Society* **8**, pp. 912–922 (1957).

[61] H. Seifert and W. Threlfall, *Seifert and Threlfall: A textbook of topology, Pure and Applied Mathematics*, Vol. 89. Academic Press (1980), translated from the German edition of 1934 by Michael A. Goldman, With a preface by Joan S. Birman, With "Topology of 3-dimensional fibered spaces" by Seifert, Translated from the German by Wolfgang Heil.

[62] A. Shima, Immersions from the 2-sphere to the 3-sphere with only two triple points, *Sūrikaisekikenkyūsho Kōkyūroku* **1006**, pp. 146–160 (1997), Topology of real singularities and related topics (Japanese) (Kyoto, 1997).

[63] S. Smale, A classification of immersions of the two-sphere, *Transactions of the American Mathematical Society* **90**, pp. 281–290 (1958).

[64] S. Smale, Diffeomorphisms of the 2-sphere, *Proceedings of the American Mathematical Society* **10**, 4, pp. 621–626 (1959).

[65] V. G. Turaev and O. Y. Viro, State sum invariants of 3-manifolds and quantum $6j$-symbols, *Topology* **31**, pp. 865–902 (1992).

[66] A. Verona, *Stratified mappings—structure and triangulability, Lecture Notes in Mathematics*, Vol. 1102. Springer (1984).

[67] R. Vigara, A new proof of a theorem of J. M. Montesinos, *Journal of Mathematical Sciences-University of Tokyo* **11**, 3, pp. 325–351 (2004).

[68] R. Vigara, *Representación de 3-variedades por esferas de Dehn rellenantes*, Ph.D. thesis, Universidad Nacional de Educación a Distancia (2006).

[69] R. Vigara, A set of moves for Johansson representation of 3-manifolds, *Fundamenta Mathematicae* **190**, pp. 245–288 (2006).

[70] R. Vigara, Lifting filling Dehn spheres, *Journal of Knot Theory and its Ramifications* **21**, 8, pp. 1250082, 7 (2012).

[71] J. H. C. Whitehead, Simplicial spaces, nuclei and m-groups, *Proceedings London Mathematical Society* **45**, pp. 243–327 (1939).

[72] J. H. C. Whitehead, On C^1-complexes, *Annals of Mathematics* **41**, pp. 809–824 (1940).

[73] J. H. C. Whitehead, Combinatorial homotopy I, *Bulletin of the American Mathematical Society* **55**, pp. 213–245 (1949).

[74] J. H. C. Whitehead, Manifolds with transverse fields in euclidean space, *Annals of Mathematics* **73**, pp. 154–212 (1961).

[75] T. Yashiro, Immersed surfaces and their lifts, *New Zealand Journal of Mathematics* **30**, 2, pp. 197–210 (2001).

Index

SERIES ON KNOTS AND EVERYTHING

Editor-in-charge: Louis H. Kauffman *(Univ. of Illinois, Chicago)*

The Series on Knots and Everything: is a book series polarized around the theory of knots. Volume 1 in the series is Louis H Kauffman's Knots and Physics.

One purpose of this series is to continue the exploration of many of the themes indicated in Volume 1. These themes reach out beyond knot theory into physics, mathematics, logic, linguistics, philosophy, biology and practical experience. All of these outreaches have relations with knot theory when knot theory is regarded as a pivot or meeting place for apparently separate ideas. Knots act as such a pivotal place. We do not fully understand why this is so. The series represents stages in the exploration of this nexus.

Details of the titles in this series to date give a picture of the enterprise.

Published*:

*The complete list of the published volumes in the series can also be found at
http://www.worldscientific.com/series/skae

Printed in the United States
by Baker & Taylor Publisher Services

Printed in the United States
By Bookmasters